L.A.W.S OF INTERIOR PLANT CARE

DESIGN WITH INTENTION · ZERO PLANT REPLACEMENTS

This manual was created in 1985 as a complete course of indoor plant care for professionals, the results of a **10-year company study of 70,000+ interior plant**s throughout the first major foliage plant boom, during the 1970s-80s.

Maintenance employees on the project kept detailed weekly reports on each of the plants in a wide variety of commercial plant locations—shopping malls, restaurants, banks, office towers, and all of the extremes of likely indoor environments that we could locate. The design department monitored their findings. After five years, the plants flourished and a **company-wide Zero Plant Replacement Rate** was achieved.

During the next five years, our designers continuously tested new findings to seek out the limitations of tropical plants indoors. We perfected techniques that allow plants to grow in the lowest possible light levels, using our newly integrated **Light • Air • Water • System (LAWS)** for indoor plant design and care.

Following the second major plant boom of 2020, during the COVID-19 pandemic, these extraordinary findings are being made available to the general public through the release of three Hamilton books—**whereby professionals and all people who enjoy caring for plants can, with confidence, quickly and efficiently take the necessary steps to ensure that their designs and plants will live on indefinitely**—to clean our air, to temper global warming, and to enhance and enrich our lives and indoor spaces.

TABLE OF CONTENTS

While not all indoor plants are included in this teaching manual,
When you learn the **basics of plant care** given here
You will know how to care for any plant just by examining its leaves.

NASA Air Purifying Plants

Hamilton's Maintenance Techniques for Interior Plants
LAWS – Light • Air • Water • System of Plant Care and Design

by David L. Hamilton and Patricia Hamilton

First Edition 1985—ISBN 0930129078
Reprint 2021—ISBN 9781953120212
Also available on-line as Hip-Pocket Edition and E-book
©2021 Patricia Hamilton
Plant illustrations Karen Wall
ALL RIGHTS RESERVED • PRINTED IN USA
www.parkplacepublications.com • Pacific Grove, California

INTRODUCTION
LIGHT • AIR • WATER • SYSTEM
LAWS of Interior Plant Design and Care*

As soon as you finish reading this book, you will know 95% of the secrets to indoor plant care. The remaining 5% will become obvious after you have watched how your plants respond to minor variations in environmental conditions—specifically light, air, and water—over the course of a calendar year. You will enjoy the challenges as enumerated herein, and the successes, as your plants grow and flourish as never before.

Indoor plants can be relatively easy. They respond primarily to light and water. And after you have selected, purchased and prepared your plants for the light levels where you wish to place them—the maintenance techniques of our Light • Air • Water • System make watering your plants an exact science. *"How much and how often"* details are provided.

Proper watering means that plants never need replacing through careless inattention. Sometimes you may mis-water a plant, but one mistake is rarely fatal. If anyone points out that "your plant is dying" (because it has 2 or 3 yellow leaves), give full attention to the perceived problem and, using the very detailed instructions in this manual, your plant will soon be flourishing and your reputation remain intact!

It cannot be stressed too strongly that watering the roots is the key to your success with indoor plants. Every time you water, you must strive to give the roots exactly as much as the plant needs. Too much or too little water causes the plant to grow less optimally.

You will notice that when your plants are growing strongly, they rarely need any trimming. Therefore, when you water accurately, your maintenance time is less, leaving you more time to enjoy your plants.

To make learning easier, the sections in the front of this book (preceding the actual plant information), have been arranged to read in consecutive order. Because all the physical and environmental influences on a plant are interrelated, you must understand all of them to get a clear picture of the best way to make a plant grow and flourish indefinitely.

—David L. and Patricia Hamilton
January 2021

• While not all indoor plants are included in this teaching manual, when you learn the basics of plant care given here you will know how to care for any plant just by examining its leaves.

LIGHT

Light enables a plant to manufacture food in the leaves. Therefore, higher light means more food production, resulting in faster growth. An indoor plant cannot get too much light. However, too much hot, direct sunlight may scorch delicate plant tissues.

It would be wonderful if all plants could sit next to a window, but at least 50% of interior spaces are lit by fluorescent light. All plants have a minimum light level. **Trees and Ferns generally need the highest light, while most Palms and Canes tolerate the least.** As you make your maintenance rounds, you will notice that a plant sitting under a bank of overhead lights has twice as many leaves as one in a dark corner—and they can both be healthy.

An indoor plant is considered happy when it grows new leaves at least as fast as it sheds its oldest leaves. When it loses leaves faster than it grows new ones—it is dying! With a little practice, you will be able to measure light with your eye. A dark corner, for example, may have as little as 10 footcandles of light (abbreviated 10fc) while a South window may have over 2000fc in the summer.

On the average, a brightly lit office has between 50 and 300 footcandles of light at the workspace. This is adequate light for almost all plants. Fig Trees and Ferns prefer 80fc or more, or they become somewhat sparse. A Bamboo Palm, on the other hand, gets by with as little as 30fc, and the famous Aspidistra can survive for years on 5fc.

When plants sit next to a window, they grow fastest during the summer. Sometimes they nearly stop growing during the winter. When solely lit by artificial light, they grow at the same rate throughout the year. Some plants grow faster than others—given the same amount of light. **As a rule, thick leathery leaves mean slow growth, while thin delicate leaves mean fast growth.** Also, plants with thick leathery leaves tolerate much less light than fast growers with papery-thin leaves. Even with bright window light, a Corn Plant (thick leaves) increases in size by a maximum of 50% in one year, whereas a Fern (tissue-thin leaves) can triple in size in a year.

You must, then, be very conscious of light and you will water in according to the amount of light a plant receives. In the plant section, the *'minimum'* and *'ideal light levels'* are given for each plant. Even though these levels are given in footcandles, use this rule of thumb:

WINDOW LIGHT IS ALWAYS 100FC OR MORE

A North window is 100fc – 200fc

An East window is 200fc – 400fc

A West window is 300fc – 700fc

A South window is 400fc – 1000fc

ARTIFICIAL LIGHT IS 5FC TO 150FC

Bright artificial light is about 80fc to 150fc
Average work space light is about 50fc to 80fc
Hallways and lobbies have about 20fc to 50fc
A dark corner has 5fc to 20fc

You don't have to memorize these numbers right away. Just keep studying all the plants you see. You will quickly be able to tell how much light a plant needs by watching how fast it produces new leaves. **Eventually you will see the point at which the plant refuses to grow. This is the minimum light.** You will also notice that the 'touchiest' plants need the most light (i.e. Fig Trees, False Aralias, etc.).

The light level values given in this book are somewhat arbitrary—although they are averages taken from samples of thousands of plants. They are arbitrary, especially where minimum light is concerned, because each plant is different. Identical-looking plants do not necessarily have the same number of roots. More roots mean it tolerates less light.

In addition, **the longer a plant grows indoors, the less light it needs.** The new foliage is better acclimated to interior lighting—which is much lower than nursery lighting. And some soils are better suited to low light watering. As well, the skill of the waterer is important. An expert can often maintain a plant in 30% less light than the minimum.

Therefore, the light levels you will read are designed to show you the degree of expertise that a person can achieve after a month or two of practice. If you keep your eyes open for the under and overwatering signs given in the plant section, you will quickly be able to judge light—and its effect on the amount of water you pour.

There are also *minimum ideal* light levels for plants. At this level, care is extremely easy and the plant retains much of its nursery-grown appearance. In contrast, a plant in *minimum* light takes on a new look—usually satisfactory—but with a greatly reduced foliage mass.

SUMMARY OF LIGHT

1. Light is used by a plant to produce food.
2. More light equals faster food production.
3. Window light is brighter than artificial light.
4. All plants have a minimum light level.
5. Light affects a plant's water usage.
6. Plants with papery-thin leaves grow fastest.
7. Plants with thick leathery leaves grow slowest.
8. Fast growers need the most light.
9. Slow growers tolerate the least light.
10. Light is the most important environmental element in plant care.

WATER

While you have no control over light, you have complete control over water. How much you pour and how often you pour it makes the difference between a healthy plant and a sick plant. **There is only one best way to hand water plants. You must allow them to dry out until they have nearly started to wilt.** If you don't let them dry out sufficiently each time, you will retard the growth—sometimes fatally. When you water, visualize the roots. The roots need both water and oxygen. They get oxygen when the soil dries out. If you keep the roots too wet, oxygen cannot enter the soil and the roots suffocate.

You will notice that some plants need water every week, while others can go for 3 weeks or more before they start to wilt. The fast growing plants with thin leaves need water most often, while slow growers with thick leaves take the least water. **Additionally, plants use water in direct proportion to the amount of light they receive.** A Boston Fern in average office lighting (80fc) may only use 1 pint of water a week, while the same plant in a sunny window (400fc) may require 4 times as much. How can you tell how much water to give? Practice will give you an exact answer, but generally plants in a bright sunny window can stand a thorough drenching, while plants in artificial light prefer only to have their dry soil moistened. If you overwater 2 or 3 times in a row, the plant may lose most of its leaves and have to be replaced. In low light, always water sparingly. You will never kill a plant by underwatering. It starts to wilt long before it collapses completely.

Over and underwatering symptoms always show up on the oldest leaves first, and the newest growth last. **There are two basic types of plants: 1) those that can drop their leaves, and 2) those that burn at the tips.** Palms and Cane-type plants have firmly attached leaves that burn. Tree-types can shed leaves and whole branches.

For Palms and Canes, when the tips of the leaves are brown and dry, the plant is too dry. When the tips are yellow and soft, the plant is too wet. The Tree-types are different. When they are overwatered, they shed healthy green leaves. When underwatered, their leaves turn yellow. Some degree of tip burning and shedding always happens unless the plant is growing in bright light. A serious watering error has occurred when damage shows up on the new growth. It is important to pour the water evenly over the whole rootmass; if you miswater some of the roots, some of the leaves will turn yellow or burn at the tips.

There are 3 basic ways to test the soil for moisture content. An ages-old way is to stick your finger into the soil. A contemporary way is to use a moisture meter. Since most plants dry halfway down between watering, the meter is more accurate. A third method is to visually inspect the plant, to check the moisture content of the foliage. The leaves feel softer to the touch, hanging limply as the soil moisture is depleted.

The moisture meter can be a tremendous aid for high quality plant care. It is self-teaching. Insert the probe into the soil until the needle swings over to the top of the scale to see how far down the plant has dried out. **When the plant has reached the wilting stage, check to see how far you have inserted the probe before the indicator hits the pin at the WET end of the scale.** This is the ideal moisture depth for the plant before watering. With very little practice you will easily be able to determine "the point of wilting" for each plant.

Moisture levels are given for all the plants. **Use this as a starting point.** There are many other factors that influence the **POINT OF WILTING**, such as pot size, overall health, as well as the amount of light. Each plant is different, but it will only take you a few visits to see how it responds to the water you gave it the prior week.

The moisture meter is extremely accurate and quite sensitive. The parts and wires inside are delicate—the tip of the probe generates a very small current of electricity when it comes in contact with the chemical salts suspended in the soil moisture. Sometimes the meters break when you drop them once, but most often they last for about six months.

In time, the probe wears away, the gauge gets jiggled out of place, or one of the wires breaks. It's a good habit to check the meter frequently by sticking it in your water bucket or under a tap of running water. When the needle "pins out" immediately as the tip of the probe contacts the water, it's working properly.

Insert the meter in 2 or 3 different spots to see how much the plant has dried out. Consult the plant care section of this book for the recommended drying levels. **Remember, each plant is a little different.** You will learn the exact "level of dryness" for each plant when you also observe the moisture content of the leaves while using the meter. Since you will generally do your maintenance once a week, you will sometimes be forced to water plants before they are ready. Simply give them a little less water than normal. If a plant has seriously dried out, make sure you give it twice as much water as normal. It will use the extra water quickly to replace the deficiency it has suffered.

SUMMARY OF WATERING

1. Plants use water in direct proportion to the light they receive.
2. The best time to water is at "The Point of Wilting".
3. Slow growing plants (leathery leaves) need the least water.
4. Fast growers (thin, papery leaves) require the most water.
5. Never "soak" a plant—except in bright window light.
6. Overwatering is 10 times worse than underwatering.
7. Pour the water evenly over the entire root surface.
8. Underwatering signs are yellow leaves, or brown tips.
9. Overwatering signs are shedding green leaves, or yellow tips.
10. A moisture meter is an exceptional training tool.

AIR & ROOTS

Leaves need light and carbon dioxide. Roots need water and oxygen. There are two basic root types, just as there are two basic foliage types. Fast growing plants like Fig Trees and Ferns generally have fine fibrous roots that can gather water quickly. Slow growing plants like Palms and Cane-types generally have thick fleshy roots that are capable of storing water. These roots gather water much slower.

Think of roots like a pump that is permanently switched to the ON position. **The roots will gather all the water they can get and pump it up to the leaves.** That is why a plant is so easy to overwater.

The slow growing plants with the fleshy roots are easier to overwater because the leaves of the slow growers transpire less water than the fast growers. Excess water collects at the tips of the leaves. If you have not done so before, take some plants out of their pots and inspect the roots. You will notice that very few plants are rootbound (i.e. 80% or more of the soilmass is roots). However, indoor plants prefer to be solidly rooted. They grow faster and are also more difficult to overwater. Many plants fresh from the nursery have fewer than 50% roots compared to soil. In typical indoor lighting (100fc), it often takes at least 3-5 years before roots fill the pot. Remember that a pot with relatively few roots requires less than half as much water as a pot that is rootbound.

If you use a moisture meter to check for soil moisture, you will be able to feel how many roots the plant has. The meter is harder to insert when there are a lot of roots. You might think that using the moisture meter regularly will damage the roots but the opposite is true. The meter helps to aerate the soil and lets more oxygen get to the roots. You should always insert the meter into 2 or 3 different areas of the rootmass when you are learning to use it. It will show you if water is getting to all portions of the roots. **Indoors, most of the roots are located in the bottom half of the pot, especially for plants in lower light levels.**

After you become proficient at maintenance, you will see that the foliage of a plant is little more than decoration (the food manufacturing process notwithstanding). The roots do most of the hard work. All of your talents must be focused on the roots. If you treat them correctly, the plants will never need replacing until they get so big they grow through the ceiling. It is not easy to learn everything about roots in a day or two, because there are so many combinations of soil, pot size, plant size, and light. But you must still try to visualize every set of roots you water, while watching to see how your watering affects the foliage.

Roots provide stability to the stem and they also store moisture. **You can tell if a plant is sufficiently well-rooted by wiggling the stem.** If the soil surface cracks even slightly, the plant is severely under-rooted. Sometimes Cane-type plants such as Dracaenas, Scheffleras, and False Aralias are installed before they have adequately developed root systems.

These under-rooted plants (especially in lower light levels) are your greatest concerns. And since the rootmass of a plant always dries out from the top to the bottom, there is an ever-present danger that the roots at the bottom of a pot will stay wet so long they begin to rot. Therefore, for a combination of reasons that include getting enough oxygen to the roots, dealing with under-rooted plants, plus the precise watering required for low light—you are left with only one system for watering plants.

To repeat, it means watering at "The Point of Wilting." When you do this, you will actually kill the roots at the top of the pot, but simultaneously grow new ones at the bottom. **Very few healthy indoor plants have active roots in the top 1/4 of the pot.** The exceptions include fast growers (thin leaves) that are installed in bright window light.

The opposite of watering a plant when it is ready to wilt, is giving it a little water each time. Many beginners are guilty of this error. Eventually, the lower two-thirds of the soil becomes so saturated that the roots die. Then, the only place they can grow is in the upper portion of the soilmass that dries out slightly each week. The result is a sickly plant that often topples out of the pot from the weight of the foliage.

The worst way to treat the roots is to give them a good soak every time. This method will make sure that the plant dies very quickly. Of course, fast growing plants in bright light may need a good soak every week, but this is an exception. Only a small percentage of the plants in interior spaces ever get that much light.

Hopefully, you will see how your maintenance is geared toward growing new leaves and roots, while shedding some of the older leaves and roots that you deliberately kill, your plants will be the healthiest they can be. It may seem to you that all your efforts should be geared toward keeping ALL the roots and leaves. Don't try it! The plants will be less healthy and will fail eventually. Remember that all plants (indoors and out) are in a constant process of growing and dying. **You must concentrate on the new growth. When it is fresh, green and vigorous—the plant is at its healthiest.**

SUMMARY OF AIR & ROOTS

1. Roots need oxygen in addition to water.
2. Roots get oxygen as the soil dries out between watering.
3. Roots will pump as much water as you give them.
4. Rootbound plants are the easiest to water.
5. Most of the roots grow in the lower half of the pot.
6. Slow growing plants (leathery leaves) are the easiest to overwater.
7. New growth (roots and leaves) is important for overall health.

TOOLS & EQUIPMENT

MOISTURE METER If you use one, this is your most valuable tool. Make sure you check every plant. Never assume the plant is OK. The more you use the meter, the better you get at maintenance. Be careful to keep the probe clean or you may get dirt on the floor. Keep a spare meter in your vehicle. Nothing is more frustrating than having a meter quit partway through the day.

WATERING BUCKETS Carry two if you can to speed up maintenance. Keep them clean on the outside. Always set them down next to a wall or a fixed object, or someone may trip over them. Fill them as full as realistically possible to cut down on the steps you take each day.

SCISSORS You will use them constantly. The sap in the plants gums up the blades, so clean them frequently. Keep them handy—a back pocket or tool belt is best. Heavy cutting may require a pair of pruners.

FEATHER DUSTER Use it gently on the plants. The dust is easily dislodged. Never dust plants that are suspected of having pests. You should spray your duster with pesticide frequently—this reduces the chances of transferring mites from one plant to another. Customers enjoy having their plants dusted. It's the most dramatic thing that you do.

INSECTICIDE Your company may have a spraying policy. If you have to spray chemicals, do it outside. Coat both the top and the bottom of the leaves—and the top of the soil—thoroughly. The insecticide has to touch the pests to be effective. If just a few bugs live through the spray, they will come back to haunt you. Insecticides are seriously poisonous. Rubber gloves and a respirator are mandatory. Wash immediately if any spray touches your skin. Insecticidal soap is safe and effective.

A SOFT CLOTH Wipe the pots occasionally. They sometimes are dirtied by the cleaning staff or through contact with shoes, etc. Rinse the cloth in your bucket to keep it clean. When "shining" a plant, rinse out the cloth every minute or two.

FERTILIZER Your company may have its own fertilizing policy. If you feed the plants, remember they only need 1 or 2 applications yearly in low light, and 3 or 4 applications in bright light. More is unnecessary (unless your company feeds a dilute solution with every watering). The best plant foods are those like 20-20-20, which contain trace elements.

YOUR HANDS AND CLOTHES Wear tasteful, casual clothes, good shoes, smile often, and wash your hands at least once on every account.

TRANSPORTING PLANTS

Whether you are loading plants on a truck or moving them from place to place on an account, there are a few simple precautions to take. The foliage (especially of leathery-leaved plants) is brittle and will crack when bent at more than a 90° angle. The easiest way to move large plants is by holding the container with both hands and carrying it at your side with the foliage trailing behind you.

If you go through a narrow doorway, look at the plant to see which way the foliage grows. Palms and Cane-type plants have downward curving leaves, so they should go through the narrow space *head first.* Plants with branches (Fig Trees, Scheffleras) are best taken *pot first.* Always be careful of the fresh new growth at the top.

Carrying plants in a van or truck is easy when you remember that the pots must not shift when the vehicle is in motion. Wedge the containers tightly together using fillers like blankets, boards, or foam to fill up the empty spaces. In a van, where plants taller than 4' must be placed on their sides, simply prop the stems on a suitable support (such as a decorative pot), to keep the leaves from dragging on the floor. **Make sure the pots won't shift, even if the vehicle stops suddenly.**

Extremes in temperatures are also a danger to plants. Even a 5 minute exposure to full summer sun can scorch the leaves, and freezing temperatures will damage the foliage in less than a minute. If the vehicle is uninsulated, the foliage will burn both from hot and cold metal. Plants left inside a vehicle with the windows closed in summer can wilt completely in 4-6 hours. Think of the plants as if they were people with sensitive and brittle skin.

You will find the plants fresh from the nursery are frequently under-rooted. Thus, if you grab them by the canes or stems, they might come out of the pot. After a while, you will see how much handling each plant can take. Of particular note are Ficus benjamina. They have a thin skin that easily becomes torn when handled roughly.

Most hanging baskets are breakable because the stems are small and often brittle. When you set them down, make sure that none of the foliage is trapped under the pot. Even removing the wire hangers from a hanging basket can be tricky when the plant is full and lush.

Bringing a truckload of new plants into a building can be a complicated operation when you are faced with elevators and a crowded loading dock. The best procedure is to unload all the plants at once, then take them all to the largest open space nearest their destination. Now the hardest job will be done and you will be able to place them in the correct spot at your leisure. If you should happen to damage a plant physically, it will probably not be very noticeable. (An exception is snapping the head from the brittle *Dracaena marginata*). When you do it in front of a customer, they are likely to get upset. Use caution even when in a hurry.

MAINTENANCE TECHNIQUES

Your company may require you to water, trim, dust and clean up at a dozen or more separate accounts during each working day. (Some companies that specialize in larger accounts often have separate personnel for watering and trimming.) There will be a fixed amount of traveling time between each account, but once you get there, be as efficient as possible. In your work, TIME is definitely MONEY.

Assuming that an average account has about 50 plants, you will find that if you work quickly you will be able to do all the service work in not more than one hour per week—not counting driving time. Some weeks you may spend additional time with sprays and cleaning, but you should budget yourself approximately 1 hour for the work. The actual watering of 50 plants (including getting the water) takes only 15 to 20 minutes.

Assuming an average of 8 plants to the bucket, you will need 6 buckets of water to service 50 plants. Take two buckets with you and fill them at the same time, setting the extra full one down where you estimate you will be when the first bucket runs out. You will notice that no matter how fast you get the water, it will take longer to get fresh water than it takes to empty the buckets. As you water, pick off the yellow leaves you see, but leave the trimming and cleaning until later (within reason). After watering all the plants, you should still have at least 1/2 hour left for trimming, cleaning, and chatting with the clients.

Trimming, cleaning, and dusting are three very separate motions. You may be required to dust every week, or sometimes only every four weeks, depending on the amount of dust in the building. Dust each plant quickly as you trim, and disinfect your duster regularly.

Cleaning is necessary for some plants more than others (especially broad, shiny-leaved plants like the Dracaenas, which show accumulated grime faster than small-leaved plants). Again, it is not necessary to polish each leaf—just do the most noticeable ones at eye-level if you are in a hurry, or all the outward facing ones when you have more time.

Trimming should be done thoroughly each time. If you do not cut off all the bad leaves, brown and yellow tips, and dead material on each visit, you will have no way of knowing how accurately you watered the previous week. **The trimming you do is your only guide to your success with plants.** The less you need to trim, the better you are at watering. After awhile, trimming should take up no more than 5 minutes of each hour. In the beginning when you are learning how to water, it may consume 30 minutes or more. When you become a maintenance whiz, you will be able to trim so carefully that you do not cut into the green part of the leaf. This is important because cut edges of a green leaf turn brown. Sharp scissors minimize the tissue damage and degree of burning. Make it a point to cosmetically shape the plants as you trim.

CUSTOMER RELATIONS

Your employer will ask you to balance the time you spend at maintenance with your customer's need to have verbal communication with you. In some offices, everyone wants to chat and it becomes an art to keep exchanges brief. You will find that a 2 or 3 minute conversation (with one or two people) is all you can allow—and still complete the maintenance work within your time frame. It's a fine balance between working and talking. You are a stranger in an office full of people and you must acknowledge them without being abrupt. But you also must be careful not have your visit turn into a social event.

How can you avoid unnecessary conversation? There is an effective method—one that combines the maintenance techniques on the previous page with training your customers on how you schedule your work. By using this system, you will P.R. the greatest number of people for your employer's dollar, while leaving the easiest work for the last. (Watering is physically and mentally taxing—compared with trimming and cleaning.)

Thus, as you make your rounds to do the watering and noticeable trimming, you will see where the largest groups of people are congregated. Later, as you go around to fine tune the trimming and cleaning, you can stop in these areas and dispense your words of wisdom. Not only will the maximum of people be within earshot, but they will see you pampering the plants. **Pick the most visible plant in the area and shine it up to perfection.** Let everyone witness your best work.

Also, make sure you P.R. the Presidents and CEO's—when you can find them in their offices! They are ultimately responsible for hiring your company. The upper management at any account is invariably friendly. Often they will mention what may be bothering them concerning the plants. They will say things that the lower echelons of workers will not. The upper management, the office manager, and the receptionist are your most important contacts.

Sometimes, you will find office people who have an active dislike for plants. Cater to them. Take the plants out of their area. Eventually, you will turn them around—or at least they will stop being nasty. You cannot afford to have a single person dislike you. They may get on your case, but you don't have to be paranoid or overly accommodating to the point of denigrating yourself. Hardly anyone dislikes the 'plant person'.

If you should have a personal confrontation with any customer that looks as if it might lead to further problems, report it to your supervisor. Most of the time the misunderstanding is easily cleared up. Remember, customers expect you to be cheerful, witty, and a pleasure to see. Large corporations have a motto which says, **"The Customer is Always Right."** Don't forget it. Follow the motto. You will always come out ahead when you are courteous and respectful.

ACCLIMATIZATION

When the light plants receive changes in value, plants make changes in their foliage and root structure. The most dramatic changes occur when plants move from a greenhouse to a poorly-lit indoor location. **An average greenhouse has 2000 footcandles of light or more, whereas the average interior space rarely has 100 footcandles.** Amazingly, plants make the adjustment rather easily. It takes just 4-8 weeks in most cases. During this period, the plants invariably shed both roots and leaves. Fast growers, like Fig Trees and Ferns, do the most shedding, while Palms and Canes suffer the least.

The most important concern for interior landscapers, is how to water during this period of adjustment. **First, you must remember that plants store tremendous amounts of food during their time at the greenhouse.** And it takes them two months or more to use up the excess food after they arrive indoors. This means they gradually use less and less water. A 5' Fig Tree may need more than one-half gallon of water during its first week indoors but only one quarter a week after it has been indoors for several months (in 100fc).

The second most important aspect of acclimatization is shedding. Fig Trees will shed up to 20% of their leaves (the oldest, interior leaves), while Cane-type plants shed their lowest leaves (also the oldest leaves). Unless you're aware that acclimatization is a natural process, you may think the shedding is your fault, because it looks like underwatering.

After you gain more experience, you will learn exactly how many leaves each type of plant sheds as it adjusts. You will remove these leaves before they have a chance to turn yellow—thus avoiding the inevitable comment your customers always make when a plant has a few bad leaves. They will all say, ***"That plant over there is dying."***

Plants also acclimatize when they are moved from an office area of low light to a window. The process is the reverse of above. They gradually start to use more and more water. This reverse form of acclimatization occurs over a longer period (about 2-4 months). As the plant orients itself to a brighter light, many of the older leaves will not be able to stand the increase in light. They will show overwatering vs. the underwatering symptoms in the greenhouse-to-indoor acclimatization.

Fast growing plants show the most dramatic change in the new light. Their leaves and roots are sensitive to minute environmental changes, while the hardier, slower growing plants always give you the least trouble. After the initial acclimatization is over, plants will settle into a routine where water requirements stay much the same for months or years. At least 50% of all plant failures occur (or are precipitated) during the first 2 or 3 months indoors. **It is a good policy to underwater slightly.** Underwatering sheds the leaves that will die sooner or later. But, if you overwater, you will cause irreparable damage to roots.

SHEDDING & BURNING

We have seen how plants as they grow are in a constant process of shedding old foliage as they grow new. Plants with leaves (Tree-types) generally shed an entire leaf at once. **The plants that make the most work for you are the Palms and Cane-types.** Why? Because the fronds (or leaves) are large and firmly attached to the stem and they gradually burn from the tip inward. It takes an average of 2 to 6 months for the frond (or leaf) to burn all the way back to the stem.

The consideration here is how to save maintenance time. It can take 5 or 10 minutes to trim up a large Palm or Corn Plant every week. The best solution is removing the entire leaf. Often, 20% of the leaves of a plant can be removed without affecting the overall appearance. Some plant companies believe in spending the time to trim, while others can see the overall economics in being more ruthless.

Here is an option you may consider: When a Palm or Corn Plant is installed, you can remove 10-20% of the oldest fronds or leaves. This will stimulate the plant to produce new growth faster since there is now a greater proportion of roots to foliage. **New growth is always the most beautiful, and it never burns!**

It is esthetically displeasing to see a section of leaves cut back into stubs. And, while I will always agree that even the tiniest speck of black, brown, or yellow on a plant detracts from its appearance, so do chopped up fronds. A small degree of trimming is acceptable—perhaps the outer inch or two of a leaf. But when a leaf must be cut back 6 inches or more, it only shows a careless lack of maintenance skill.

As well, the customer rarely knows how a plant looks when it is 'nursery-fresh'. Instead, they would rather have a plant that looks as if it is doing well. We cannot fool the client with severely cut back foliage. Also, constant trimming seems to offend many lay persons. They cringe when they see us attack a plant with scissors.

Plants have a wonderful way of filling in their bare spots. We know that plants orient their foliage to get a maximum of light. **When some fronds or leaves are cut off, others will move in to take their place.** All plants grow symmetrically when left to their own devices. Arguably, the family of Dracaenas (the backbone of our industry), may look somewhat leggy when a 6-8 footer has no leaves at all on the lower 3 or 4 feet. In that case, perhaps the solution is to replace the plant.

If you remove the occasional plant from the job site, when it is unattractive to you—but still acceptable to the customer—they will appreciate your high standards. And when a plant is chopped up through poor placement or watering techniques, it is best to admit your error by giving the client a fresh one. Often you can try to save a few dollars in replacement costs but lose the account through obviously shoddy work.

HEAT & COLD

Generally, tropical indoor plants respond favorably to temperatures between 68°F and 86°F. From 62° to 65°, there is a noticeable slowing in growth, while temperature over 78° can cause rapid wilting—especially in low humidity. Fortunately, office buildings keep their thermostats between 68°F and 72°F, which is the perfect range for indoor container plants. Sometimes you will maintain plants that sit in lobbies subject to blasts of cold winter air, or sit next to windows with poor air circulation in hot summer months. Here is what you do.

In the cool lobby, you will notice that a plant may take 3 or 4 times as long to dry out as normal. Water it as sparingly as possible. Keeping the rootball too wet can cause massive shedding in fast growing Tree-types. Since the growth rate is slowed so much, the plant may never recover from a single overwatering.

When the temperature is 60°F or less, some durable plants, like the Yucca, may not need water for 3 months or more. **You can always tell which plants can stand the cold best, because they are the slow growers with thick leathery leaves.** The intensity of light also makes a difference to plants growing in cool temperatures. When the light is bright, the plants are much less apt to suffer from overwatering.

Hot windows or atriums are less of a danger than the cold, although you may find that your maintenance has to be performed twice a week rather than once. If you run across a situation where no matter how thoroughly you soak a plant, it is dried out before you come back a week later—there is a simple solution that works every time. **Put a plastic saucer 2 inches deep under the nursery container** and make sure it's full of water after you water the plant. The plant will exhaust the water in the rootmass and draw up the water from the saucer like a wick. There will be plenty of water for 7 days. This procedure works because it is impossible to overwater a plant in a hot, bright window.

Alternatively, you can repot into a larger container, and put extra soil on top of the rootmass. Or put 2 or 3 inches of bark mulch on the top of the rootmass. Or you can retrofit with a subirrigation system. All of these solutions will prevent the roots from drying out too fast. In hot, bright places, plants grow surface roots, whereas in low light, most of the roots grow toward the bottom of the pot. So, the addition of a thick layer of mulch prevents rapid drying of the upper soilmass.

Plants growing next to a sunny window have a nasty habit of pressing their leaves against the glass. The heat and bright light discolor them quickly, and since the effect of scorched leaves looks terrible from outside the building, you must keep the plants trimmed well back from the glass. You can also deal with the problem by giving the plants a 1/4 turn every few weeks. However, sometimes the plants will be planted-out directly into the soil and your alternatives are limited.

HUMIDITY & MISTING

The humidity of interior spaces is really of minor concern to you, because most buildings are rigorously climate-controlled. Sometimes, however, you are faced with the dryness of forced-air heating or the wetness of sticky summer weather. The summer humidity is a blessing because tropicals absolutely love high humidity. It makes them use much less water and also helps them retain more foliage, since the rootmass dries out less between waterings.

Dry air is a major enemy of container indoor plants and it affects the thin-leaved, fast growers more than the slow growers with their thick, leathery leaves. **Dry air can cause the leaves of a plant to transpire moisture faster than the roots can pump it up.** Fortunately, low humidity is rare.

You will be asked one question more than others when you go on your maintenance rounds. ***"Does it help to mist my plants?"*** This is a loaded question because those who ask it often take great pride in their plants and they mist them frequently. People who believe in misting their plants are often those who like to water them as often as possible.

Misting the plants has long been a basic part of homeowner plant care. It is a principle that is repeated in almost every houseplant book. It ranks up there with instructions to keep every plant "evenly moist." But what does misting do? It raises the humidity for as long as it takes the moisture to evaporate from the foliage.

If you have ever misted, you will notice that the droplets of water evaporate in 30 minutes or less. **As a result, the misting helps to the extent of raising the humidity while the moisture stays on the plant.** It follows then that for misting to be completely effective, the plant would have to be misted every hour or two. Yet, because most people rarely even mist once a day, you can see that misting is generally of little benefit at all. During the summer, however, when spider mites multiply the most rapidly, misting a plant can go a long way toward keeping the mites from running completely wild. But, there is a negative aspect to misting, especially on plants with glossy leaves.

Water contains minerals that spot the foliage after it evaporates. So, if a plant is well shined and dusted, the residue from misting will quickly mark the surfaces of the foliage.

Standing the plants on a pebble tray that is constantly filled with water is more effective. As long as the pot holding the plant is elevated above the water, the evaporation of the water in the saucer underneath will bring a constant increase in humidity.

Misting is an age-old custom. People like to do it. However, don't do it to the plants on your maintenance accounts. It is unnecessary commercially, and you can use your service time more productively.

CLEANING

Plants with shiny, dust-free leaf surfaces look healthy and attractive. Plants should be just as clean as the rest of the furnishings in a room. You will find that many of your customers, especially if they are not overly fond of plants in the first place, will be quick to mention what they consider is a "dirty plant."

At the nursery, tropical plants are sprayed with insecticide on a regular basis. As a result, the foliage is heavily overlaid with chemical residues plus minerals dissolved in the water. Sometimes the plants are thoroughly cleaned before they leave the nursery, but often you may be responsible for doing the final cosmetic cleaning of the leaves. **Cleaning is more of a science than you may imagine.** First, you are trying to put the best possible 'finish' on the leaf surface, and second, you must do this without damaging the plant.

Your employer will probably supply you with one of the excellent commercial leaf cleaners and/or polishes manufactured for this purpose. They cut through the grime and leave either a natural, waxy, or high gloss shine on the leaf. The type of finish you decide to put on the leaf depends on your personal tastes. Some people prefer the natural look of a plant's own waxy foliage coating. Others desire a light shine. And some people enjoy seeing leaves that are bright and shiny. Whichever cleaner or polish you choose, it should be used two or three times a year, at most. The rest of the time, dusting will keep your plants clean. **It would be best if you never had to clean a plant by hand**, because plants don't appreciate a lot of rubbing on their leaf surfaces. If you do get a coating of polish on the undersides, you can cause tissue damage by plugging the pores that exchange oxygen and carbon dioxide.

There are two mechanical considerations in applying any cleaner which contains a waxy or oily polish. First you must keep the polish away from the bottom of the leaf. You can do this by spraying from the top, or by tipping the plant forward and spraying perpendicular to the leaf surface. Secondly, after the product is applied, you must rub evenly. When you 'polish' the leaves, you must rub lightly and **take extreme care not to crack the tissues** or dislodge the leaves. (The Dracaena deremensis 'Warneckei' is one of the most difficult plants to clean without pulling off the leaves.)

On a plant 5' tall, you will find it necessary to rinse your cloth or sponge five to ten times as it picks up the dirt. There is a distinctly separate coordination of your two hands for each type of plant. Largely, you will find yourself supporting the underside of the leaf with one hand as you rub the top with the cloth.

When you clean Palms, especially thin-leaved ones like the Bamboo Palms, you may need to cut the recommended dilution in half as they can't stand a heavy surface coating of polish.

PRUNING

Most of the Tree-type plants need occasional cutting back to keep them under control. You will also have to cut back many of the hanging Vines to keep them from getting too leggy. You will never cut back any of the Palms unless the entire stem is too tall and you want to remove it completely.

When you prune a Tree, Vine or Cane, the plant will sprout new growth underneath the cut. **The number of new heads or branches that grow depend on the light intensity and the plant's overall vigor.** Plants fresh from the nursery sprout new growth faster because they are more vigorous. An older plant in relatively low light may wait 4-6 months before it gradually starts to grow new leaves near the cut.

You can tell which way the new branch will grow by looking closely at the leaf underneath the cut. The new sprout will grow directly from the point where the leaf joins the stem. The easiest plants to prune are Fig Trees. They grow rapidly and produce new branches in profusion.

Basically, when you cut back a Tree-type plant (including Scheffleras, False Aralias, etc.), you must decide where you want the new branches positioned. Sometimes you may cut the existing branch or stem off at the base. As you do your maintenance, look carefully at how the plants have been pruned in the past. This will give you clues to what you can expect when you prune.

You will probably cut back the vines most often (especially Pothos, Grape Ivy, and the many varieties of Philodendrons). **Generally, you cut them back hard** (i.e. close to the soil level). This is because the stems are often too weak to support tip-growth branches. You prune Vines for two reasons—to eliminate gangly growth, and to force the plant to produce its foliage in a symmetrical fashion close to the pot.

The most fun plants to prune are the Cane-types—especially the Dracaena marginata. These plants are produced at the nursery as a multiple collection of single stalks in the pot. The stalks simply grow taller and taller, never developing side branches (except under special circumstances). You must be careful not to prune them when they are sparse to begin with, because it takes six months or more to fully grow new sprouts.

Always use a sharp pair of pruners to cut a plant back (except Vines) because scissors will not be strong enough to make a clean cut. **Try to do your serious pruning when there are no witnesses.** Pruning is the one maintenance operation that the lay person disapproves of universally. Do a good job of cleaning up after you finish pruning.

You will not have to do much pruning except on Vines and some Trees. Pruning is a job for the experienced—when in doubt consult your maintenance supervisor first. **Often a plant that looks poorly in its space may look good somewhere else.**

PROPAGATING

You will seldom be called upon to actually propagate a plant, but your customers will often ask you how it's done. Also, when you cut back a plant, especially a Vine, you will be asked for the cutting. It helps your credibility to know whether the cutting will take or not.

Many of the plants you care for were grown from cuttings at the nursery. **Nurserymen have the benefits of bright light and high humidity** to help them grow roots on the section of the plant they are propagating. Even in the nursery, propagating is not an easy job. Often the cuttings are placed under a plastic tent to encourage rapid root growth—before the foliage wilts from lack of internal moisture.

Plants with the fleshiest stems and leaves are the most straightforward to propagate. This makes Succulents the easiest and woody Tree-types the most difficult. Vines, such as the Pothos, are somewhere in the middle. A 4" to 6" cutting from a healthy Fig Tree branch will readily root in water, when it is given bright light, humidity and warmth. Propagate a few Fig cuttings in one jar to later create a very attractive bonsai planter.

Most office people want to put the cuttings in water and this is the easiest way to preserve them—at least temporarily. If they should happen to put the cuttings in water, next to a window, and change the water on a weekly basis, the plants may survive for six months or more. Most people who ask for cuttings will not ask for advice on how to grow them. They likely will have done it before. The only question you may encounter is how to plant the cutting in soil after it has grown roots in the glass of water.

The plant will have grown 'water roots,' which are different from 'soil roots.' **The water roots may die quickly when they are placed in the confines of soil in a pot.** The procedure may only work a small percentage of the time—usually when the cutting is small and gets plenty of light, warmth and humidity.

Succulents are very easy and can be grown from a single leaf or a stem section. Because they have so much internal moisture, plus a thick skin that prevents moisture loss to the atmosphere, they will grow new roots most of the time. **They must be calloused first** (i.e. left exposed to the air until a protective skin forms over the cut part). Otherwise they will rot.

There is a single important rule. The soil must be dry. Never water until after the roots have grown (often two months or more). If you use even slightly moistened soil, the cutting will rot rapidly. It's obvious that the soil doesn't need any moisture because there are no roots to moisten. The cutting produces roots from the moisture it has stored in the stems and leaves. A pebble tray helps to provide humidity to the leaves until the roots grow. Water the tiny new roots sparingly as they grow. Usually a few dribbles around the stem is sufficient.

REPOTTING

Since one purpose of this book is to examine the many myths on plant care, the subject of repotting is the best example of how not to follow the popular concept of dealing with a plant's roots. Even in the present day of sophisticated plant care, you may be under the impression that a container plant's roots need plenty of room to grow. *Nothing could be farther from the truth!*

You may believe that when roots grow out the bottom of a nursery container, it is time to repot. Roots will always grow out of holes in a pot—especially if there is drainage moisture in the decorative container. **Roots are constantly searching for moisture and they will grow anywhere they can find some!** You may have heard that when the roots of a plant wind around the sides of the pot, the plant should be repotted, but this is also a fallacy. Plants in clay pots always grow roots next to the porous surface, and any plant, as it grows, naturally fills the edges of the pot as well.

Also, you may have been told that when you see roots at the top of the soil, you should repot. Again, there is no basis for this theory. **The only plants that grow roots at the top are fast growers in bright light.** An indoor plant always grows its roots outward and downward.

When is it time to repot? The answer is *"almost never."* We have already seen in the section on roots that tropicals prefer to be rootbound. The more roots, the easier they are to water, and the healthier and faster they will grow. On the average, a normal 5' plant in a 10" pot has enough space in the nursery container for 3 to 5 years of root growth. Fast growers, of course, may need repotting more often.

There is a very easy method for repotting. Whenever you inspect a root system and see that the bottom of the soilball is solid with roots, you can add 1 or 2 inches of new soil at the bottom of the pot and replace the soilball. This will be enough root space for a few more years of growth. Scrape the excess of unrooted soil from the surface of the rootmass. This way you can keep the same plant in the same pot for years. Because nursery containers are slightly tapered, you will be able to add one inch of soil all around the sides of the pot. Tamp the soil at the sides down tightly and eventually new roots will grow there as well.

When a plant is repotted, loosen some of the roots before you put the soil back in the pot (both the bottom and the sides). This helps encourage root migration. Roots grow more slowly indoors and the inch or two of new soil will last much longer than you expect. It is possible (in low light) to keep a plant in the same pot for 10 years or more, even if it grows from a 5 footer into a 10 footer. Transplanting into a large pot is only necessary when the roots are drying out faster than once a week even when you soak them.

SOIL

A good knowledge of soil is important because the various types accept water differently. There used to be three major rooting mediums. One not still common is sandy. The soil can be wonderful for the nurserymen, because it accepts water and drains readily. It allows nursery plants to grow quickly. It is disadvantageous indoors because it is fairly cold, and it shrinks noticeably as the plant dries down from its nursery 'evenly moist' condition. Correct the problem by packing the soil down tightly around the edges of the pot—twice during the first six months indoors. A one-half to three-quarter inch stick works best on a 10" to 14" pot. **Always tamp the soil down around the edges as soon as you take the plant from the greenhouse to the indoor location.** If you don't, an air space forms around the edge of the pot which makes watering extremely difficult. Since there is more shrinkage in low light (less soil moisture), repeat in six months. Tamp in additional soil.

Local nurseries in your area may use their own combination of peat and perlite and sand, often substituting styrofoam pellets for the perlite. You will often find this soil in hanging baskets and small table plants. It is warmer than sand, but depending on the peat content, it can form a crust on top as the peat dries out. Stir the surface soil to get rid of the crust; often the addition of a handful of sand (or perlite or foam chips) is better. This soil mix holds water much longer than the sandy mix.

Plants from Hawaii may be planted in ground-up pieces of lava rock. **This is by far the most difficult soilmass to deal with.** It is hard to measure the moisture content because the particles are so coarse (especially when the plant is under-rooted). The moisture meter or the finger thus measures less moisture than there really is. Also, it dries out faster, and can be a major problem in bright light.

You can only improve on this 'crunchy' soil with great difficulty. In reality, the plant should be replanted before going into an interior space. However, you can add sand to the top of the pot as fast as you can get it to wash down into the cracks. Eventually, the sand will be equal in volume to the lava rock and watering will be easier. Other soils can be quite loamy (most garden soil), and some soils have a high proportion of bark. All the different types of soil affect how fast a plant dries out, and how fast it grows. After a while you will be able to feel the different types of soil as you probe with a finger or moisture meter.

Which is the best soil mix for indoor plants? If you are not satisfied with the one you are using, **try one of the excellent packaged commercial products**, or try a mixture of 12% sharp sand, and 34% medium to coarse peat, and 54% coarse perlite or styrofoam. It holds water well, doesn't shrink, and drains efficiently. Also, it stays much looser as the roots are growing, and is much warmer than a mix that is predominately sand.

MULCHES

The mulch that is used to cover the surface of the soil and hide the nursery pot (inside the decorative containers), is merely a decoration—unless it is used to retain moisture in hot, bright, dry locations. Mulches commonly used include moss, bark chips, rocks and cork.

Formerly, bark was the most common, but now sphagnum moss or Spanish moss has increased in popularity. In truth, **plants are better off without a mulch** because it retards the soil from drying out. The major problem with indoor plants is the difficulty in drying them out fast enough to prevent overwatering. For this reason, rocks are by far the worst mulch. They can make the soil take twice as long to dry. Bark and moss are fairly equal. Large chunks of bark are preferable to small pieces because they create a more porous soil topping. Moss is generally good unless a coarsely-textured 'ground moss' is chosen. It tends to mat into a solid clump that restricts air flow.

Esthetically, nothing looks better than fresh green moss. Unfortunately, it rarely stays green longer than a few weeks or months. Then it fades into a dull grassy brown. The bark, at least, keeps its color throughout its life. There are suggestions that bark can change the pH of the soil, but this is a minor consideration.

Bark is difficult to work with if you use your fingers for checking the soil moisture. After digging in bark for a few hours, your fingers will be full of splinters. Moss, on the other hand, is a small problem when you use a moisture meter, since it occasionally clings to the probe and has to be repositioned. Cork is an exceptional compromise. It is manufactured in a wide range of decorator colors; it has an indefinite lifespan; and it is easy to work with. Moss can be colored as well by spray painting it green prior to installation.

No matter which mulch you are required to deal with, remember to keep it neat and tidy. This requires spreading it evenly over the surface of the soil, and making sure that it covers the nursery container it is intended to hide. Or, you can make it cover the edges of the nursery container, while leaving the center of the rootmass exposed.

The best moss/soil surface combination takes a little extra effort, but rewards you with an easy-to-water plant. To get the best results, follow the procedure for packing the soil tightly around the edges of the pot (see the section on repotting). This will prevent water from running down the edges. Then make a hollow, one inch deep in the center of the soil. Finally, spread the mulch around the edges of the pot. Now, when you water, you can see the soil surface. You will be able to fill the depression with water and be confident it will sink straight down into the roots. **After you have watered the plant 2 or 3 times, the soil will be accustomed to the new method and your maintenance times will decrease significantly.**

FERTILIZER

Since fertilizer is commonly called 'plant food', most people believe that plants like to have plenty of it. They forget that plants make most of their 'food' (i.e. energy) when they combine light with water. Plant food is really just a method for supplying minerals that may be missing in the soil. Nitrogen, phosphorous, and potassium are the most important minerals to a plant. The plant also needs dozens of minor elements. Plants store the minerals in the roots, stems and leaves, and use them as required.

In the nursery, because the soil is leached by constant watering, and because growth is so fast, plants are given relatively large amounts of fertilizer. Indoors, however, plants need as little as 1/100 as much fertilizer. In fact, **because plants store so much fertilizer in their tissues, they normally don't need any extra food at all for most of the first year** (except possibly in very bright window light). As you might assume, fast growing plants can use more food than slow growing plants. It's virtually impossible to underfeed a plant in an average indoor location—but it's easy to overfeed. When there is too much fertilizer in the soil (the concentration of salts increases as the soil dries out), the root hairs burn, and in turn, the leaves will get brown.

You will be on the safe side if you feed once a year in low light, and two or three times a year in high light. Your company may have a specific feeding program, so consult your supervisor. When you are left to your own devices, you can rely on a 20-20-20 fertilizer containing trace elements. Two teaspoons is sufficient for a gallon of water.

You may feed plants on a more regular basis if you dilute accordingly. Assuming that you water a plant weekly, divide the package recommended dosage by 1/25, and feed with every watering. This will allow for the leaching that occurs with regular watering. **Always stop feeding immediately if you see any yellow-turning-to-brown spots in the middle of the leaves.** This is a sure sign that the roots are being burned by too much fertilizer. You can also use an organic plant food with every watering. Seaweed is excellent because it contains an incredible number of trace elements. It is nearly impossible to use too much seaweed. This product also helps 'green-up' your plants within 7-10 days. Chlorosis—pale leaves with dark veins—is corrected by an application of chelated iron.

It is impossible to stress too much how important it is to be careful with plant food. **Fertilizer is not nearly as important to indoor plants as correct watering.** The plants are more concerned that you don't overwater them. (Here is a common problem with double potted plants—especially Dracaenas. They often get fertilizer burn when the roots grow out of the bottom of the pot into the drainage water. The leaves burn when the water dries up. Simply trim back the roots to prevent the problem.)

PESTS

There are 3 major pests designed to haunt the conscientious plant person. **The worst of the three is spider mites, followed by mealy bugs and then scale.** On a percentage basis, you will encounter spider mites 90% of the time you have a problem with pests. Mealy bugs comprise 8% of the problem, with scale (fortunately), only 2% of the time. All pests are prevalent during the summer. **Pests are almost always found on the undersides of the leaves.** Scale can be on the surfaces of the leaves, and mealy bugs often congregate where the leaves join the stem.

Mites are the easiest to kill. It takes a practiced eye to spot them in their early stages. **You can suspect mites when a plant suddenly stops using water.** Later the leaves become mottled. The mottling is unlike the paleness of overwatering. Since mites occur mostly on the undersides of the leaf, look for tiny white dots raised 1/16" above the surface. These are usually gathered in small clusters of up to a dozen tiny white pinpricks. These dots are actually the eggs. The clusters are usually found in the hollows formed by the wrinkles in the leaves. When the infestation is severe, you will notice webs (like miniature spider webs), and you will see the tiny mites scampering back and forth. At this stage, the plant is in serious trouble, and may be a write-off, because the mites will have already done serious damage to the leaves on which they are feeding. To kill mites, you can spray with whichever chemical your company has the best success.

You must spray every square inch of upper and lower leaf surfaces, plus all the stems, mulch and soil surface. Otherwise, the mites that survive will soon multiply. **When you spray, a commercial sticker/spreader** helps spread the pesticide evenly. (Or, in a pinch, use 1 tsp. of liquid dish soap per gallon of spray.)

Mealy bugs are small, 1/8" to 1/4" long, fluffy white insects. They spread slowly and are easily controlled by a chemical spray. However, while you can often kill the mites with a single, thorough spray, the mealy bugs may need 2 or 3 applications of the pesticide.

Scale is the most obnoxious. When the young scale hatch in spring, they succumb to chemical sprays. The adults are harder to kill. If the plant is small—and there are too many scale to scrape off by hand— throw out the plant. Larger plants can be effectively treated with a systemic that is watered into the soil or inserted underneath the bark.

You may be prevented by federal, state or local government regulations from using chemicals in public buildings. Many states now require that pesticide applicators have a license. In any case, remember that the chemicals are dangerous, and if you can, take the plant back to the greenhouse for spraying. Always use a respirator, gloves and protective clothing.

QUESTIONS & ANSWERS

1. Which *group of plants* generally have a higher minimum light requirement: Trees & Ferns or Palms & Canes?
   Answer: page 2, paragraph 2.

2. Which *type of leaves* generally has a higher light requirement: Thick leathery, or thin delicate?
   Answer: page 2, paragraph 6.

3. How do you determine a plant's minimum light?
   Answer: page 3, paragraph 2.

4. When is the correct time to water every plant?
   Answer: page 4, paragraph 2.

5. How can you tell when a plant is seriously miswatered?
   Answer: page 4, paragraph 4.

6. What characteristic of a plant's roots make it easy to overwater?
   Answer: page 6, paragraph 2.

7. Why do indoor plants prefer to be rootbound?
   Answer: page 6, paragraph 3.

8. Where are most of an indoor plant's roots located?
   Answer: page 6, paragraph 4.

9. How can you tell when a plant is sufficiently well rooted?
   Answer: page 6, paragraph 6.

10. When transporting plants, what is the prime consideration?
    Answer: page 9, paragraph 3.

11. How much time should you budget weekly to maintain 50 plants?
    Answer: page 10, paragraph 2.

12. Why should trimming be done thoroughly on each maintenance visit?
    Answer: page 10, paragraph 6.

13. How many footcandles of light does the average greenhouse have? The average office space?
    Answer: page 12, paragraph 1.

14. Which plants suffer the most shock during acclimatization? Fast growers or slow growers?
    Answer: page 12, paragraph 6.

15. What is the favorable range of temperatures for tropical plants
    Answer: page 14, paragraph 1.

16. What are 2 methods for making a plant in an extremely hot, bright location have enough water to last one week?
    Answer: page 14, paragraphs 4 & 5.

17. What major difficulty does a plant have in dry air?
Answer: page 15, paragraph 2.

18. Does misting really help? Explain.
Answer: page 15, paragraph 5.

19. How much should plant shine by diluted for thin-leaved Palms?
Answer: page 16, paragraph 6.

20. When a plant is pruned, what two things govern the number of sprouts that grow underneath the cut?
Answer: page 17, paragraph 2.

21. Why do cuttings take easily at the nursery?
Answer: page 18, paragraph 2.

22. Explain why a cutting should not be watered until it grows roots.
Answer: page 18, paragraph 7.

23. Give a reason why a plant's roots grow out of the bottom of a nursery container into a decorative pot.
Answer: page 19, paragraph 2.

24. When do the roots of an indoor plant grow at the top of the soilmass?
Answer: page 19, paragraph 3.

25. Which is the greater problem with sandy soil in an indoor situation: Cold roots or soil shrinkage?
Answer: page 20, paragraph 1.

26. Why are plants better off without a mulch to cover the soil?
Answer: page 21, paragraph 2.

27. Why do plants need very little food during their first year indoors?
Answer: page 22, paragraph 2.

28. In order of frequency of occurrence, name the 3 major plant pests.
Answer: page 23, paragraph 1.

29. Name the 8 families of indoor plants.
Answer: page 28, paragraph 1.

30. Why do you water a plant with tightly-packed roots in three or more stages?
Answer: page 30, Growth & Care.

31. Compare the differences in response to underwatering between Fig Trees and Rubber Trees.
Answer: page 31, Underwatering.

32. Which Buddhist Pine needles are affected first by too little water?
Answer: page 32, Underwatering.

33. Why is a Norfolk Pine one of the most difficult tropicals to water?
Answer: page 33, Overwatering and Underwatering.

34. What should you do with the suckers of a Schefflera in low light?
Answer: page 34, Growth & Care.

35. Name the plant that looks similar to the Schefflera, but is hardier.
Answer: page 34, Growth & Care.

36. Considering the density of the Hawaiian Schefflera's rootmass, how does it compare to the Fig Tree and the Schefflera?
Answer: page 35, How much to water.

37. Compare the color of tip-burn and overwatering on Palms fresh from the nursery to plants that have been indoors for a year or more.
Answer: page 37, Overwatering.

38. Which is the best Bamboo Palm for extremely low light: Neanthe Bella, Erumpens, or Seifrizii?
Answer: page 37, Growth & Care.

39. Why do the smaller plants in a multiple planting of Kentia Palm die back in light below 100 footcandles?
Answer: page 38, Growth & Care.

40. What makes the Areca Palm a poor commercial plant?
Answer: page 39, Growth & Care.

41. Which of the two common Cane-type plants has the highest minimum light requirement? Name the two plants.
Answer: page 41 and 43.

42. Name the 3 most popular broad-leaved Dracaenas. Compare leaves.
Answer: page 42, Growth & Care.

43. Which is the most durable of the broad-leaved Dracaenas?
Answer: page 42, Growth & Care.

44. Why does a Cane-type Dracaena need more light for the first year?
Answer: page 43, Growth & Care.

45. Which tropical plant has the most brittle head?
Answer: page 44, Growth & Care.

46. What happens to the cane of a Dracaena marginata after it has been overwatered for an extended period?
Answer: page 44, Overwatering.

47. Which is the better indoor plant: Chinese Evergreen or Dieffenbachia?
Answer: page 45, Growth & Care.

48. What should you do when a Spider Plant's leaves burn?
Answer: page 46, Overwatering.

49. How do you propagate the 'babies' produced by a Spider Plant?
Answer: page 46, Growth & Care.

50. What happens when a Philodendron's leaves are securely supported?
Answer: page 47, Growth & Care.

51. What is better than dirt as a rooting medium for Philodendrons?
Answer: page 47, Growth & Care.

52. The Nepthytis is typical of low light Vines. How can you enhance the plant's performance in light below 60 footcandles?
Answer: page 49, Growth & Care.

53. Compare the Creeping Fig and English Ivy with respect to water needs and humidity.
Answer: page 51, Growth & Care.

54. Most tropical plants have a margin-for-error in watering of 10% to 20% or more. What is the margin for error on a Prayer Plant?
Answer: page 52, Growth & Care.

55. Compare the root structures and water usage of the Asparagus Fern and the Boston Fern.
Answer: page 55, How Much to Water.

56. How long will it take for a Peace Lily's leaves to 'stand up' after they have collapsed from dry roots?
Answer: page 57, Growth & Care.

57. Most tropical plants grow taller as they age. Do Peace Lilies?
Answer: page 57, Growth & Care.

58. Which are the two hardiest members of the Lily family.
Answer: page 58.

59. How do you test a Jade Plant to see if it needs water?
Answer: page 59, How Much to Water.

60. What are the two basic physical shapes of Succulents?
Answer: page 59, Growth & Care.

61. Name the most durable type of Cactus. How long will it last in low light, providing it is not watered?
Answer: page 60, paragraph 2.

62. Assuming that you can now answer most of these questions without looking up the answers, how much do you know about plants?
Answer: page 1, paragraph 1.

THE 8 FAMILIES

Now that you have a basic understanding of the physical and environmental influences over a plant's life, you will also appreciate how much easier caring for them is to learn when you see how all indoor plants fall into 8 separate families. Here are the 8 families:

1. TREES   5. FERNS

2. PALMS   6. LILIES

3. CANES   7. SUCCULENTS

4. VINES   8. CACTI

While not all indoor plants are included here in each of these families, when you learn the basics of plant care given here you will know how to care for any plant just by examining its leaves.

The members of each of the families are similar in growth, care and watering—although their minimum light tolerances vary widely. As a rule, the Canes need the least light. The hardier Lilies are a close second. Palms are next. Then come Trees. Ferns and Vines are much the same, and lastly are the Succulents and Cacti. Except for Succulents and Cacti (which are extremely easy to water), the other families can be rated for watering difficulty by their order of light requirements. The more light a plant needs, the 'touchier' it is to water.

The most effective way to use the following plant information is to read every story. You will see a pattern unfolding that shows the minor comparisons and contrasts between plants. **Plants are basically similar,** but to the beginner each one looks different and unpredictable. Simply remember that light and water are nearly everything you have to know.

There are more commercially-hardy plants than you will find in the subsequent pages; however, you will be able to locate a representative-type of any tropical you encounter. **As you gain experience, it becomes apparent that the foliage will give you all the clues you need for care—even if you haven't seen the plant before.** Foliage can be thick and succulent, thin and succulent, thick and leathery, thick and woody, and thin and woody. The moisture content of the leaf, combined with the surface texture, tells you what the roots are like.

Tree-types with slightly succulent leaves need more light and water than those with thicker, fleshier leaves. With a Tree, the fleshier the leaf, the fleshier the root. Since a fleshy leaf uses less water, the plant tolerates less light. Some Tree-types, like the Pittosporum, have heavy woody leaves. They have the woodiest roots and need the most light.

The leaves of Palms range from grass like and thin, to thick and leathery. The thinner leaves need the most light. Just like the Trees, some of the Palms have extremely woody foliage. The Fan Palms are the best example. They need more light than Palms whose leaves are slightly succulent (Kentia and Rhapis).

The moral is to **feel the leaves of all the plants you see.** Eventually you will detect the tiny differences that will help to make you a proficient plant caretaker. There is another reason for feeling the leaves. Their moisture content tells you—even better than a moisture meter—when the roots need water. It takes years of practice to refine this technique of checking for water, but if you use your meter faithfully, the minute differences in cell moisture become apparent to you.

Of the 8 families, Vines are the most variable collection. Some of the Vines are quite succulent while others are woody like Trees. The reason they are included in a single family is because they are small and quite delicate. When Vines grow out-of-doors, they are often very large. The larger the plant, the more durable it is. It follows that indoor Vines are extremely juvenile, which makes them harder to care for.

In the family of Canes, the fleshy leaves need more light. Thus, the Dieffenbachia has a greater sensitivity to low light than the similar-looking Aglaonema. The Yucca has leaves that are woodier than the broad-leaved Dracaenas, thus it requires more light.

Succulent plants with the leatheriest leaves need the least light. Succulents and Cacti have woody roots. Thus, they are a curiosity in the indoor plant world, since succulent foliage normally means succulent roots. This combination of woody roots (which gather water quickly), and fleshy leaves (which store water easily), makes them easy to overwater. A Succulent or Cactus disintegrates when it is too wet.

Every time you see an indoor plant, concentrate on the family it belongs to and compare it to the other family members. You will soon see their strengths and weaknesses. This will give you a distinct edge when you are asked to find the best plant for a particular spot. *Interior plant design is primarily a decision on what foliage-type suits the location.* Knowing all similar foliage-types (which fall within the same family) means you have become a complete plant person.

Maintenance Techniques for the following plants include information for selection and care: botanical and trade names, family and form, standard and other sizes available, minimum light requirements, minimum ideal light requirements, level of care required, physical durability, resistance to pests, and other plants that are similar. Care signals and techniques include: When to water—by touch, with a moisture meter, and visually. How often and how much to water—in bright light and in low light. Overwatering and underwatering symptoms and how to rectify. Growth habits and quirks.

FIG TREE

Botanical name: *Ficus benjamina*
Trade name: Benjamina
Family & Form: Tree (bushy)
Standard size: 4' to 6' tall
Other sizes: From 1' to 20'+
Minimum light: 80fc (High artificial)
Minimum ideal light: 100fc+ (North window)
Care: Difficult for beginners
Physical durability: Excellent
Resistance to pests: Fairly good
Similar plants: *Ficus nitida, 'Alii'*

WHEN TO WATER: For plants in minimum light (80fc) in a 10" pot.

BY TOUCH: When soil is barely moist (slightly spongy).
MOISTURE METER: Insert 1½ ", and indicator deflects 1/2 or less.
VISUALLY: Leaves hang 45° below normal & new growth is limp.

HOW MUCH TO WATER: In light below 150fc, be careful to water only the upper 1/4 to 1/2 of the soilmass (approx. 1½ quarts every 7 days in 100fc). In bright window light (over 400fc), keep fairly moist (approx. 2½ quarts every 4 days). Be sure to water slowly and evenly.

OVERWATERING: This plant sheds healthy green leaves when the roots are even slightly too wet. Cool temperatures and drafts have the same result. If you have overwatered a Benjamina in low light, and more than 15% of the leaves have shed, move to higher light or the shedding will continue. In good light, the shedding will stop within 7 to 14 days with reduced watering.

UNDERWATERING: Masses of yellow leaves will follow even brief periods of soil dryness. When this plant moves from the nursery to a low light location (below 150fc), up to 20% of the leaves may turn yellow. Acclimatization starts after the 7th day and ends by the 21st day. Check for dry pockets of soil when there are just a few yellow leaves – otherwise soak thoroughly (one time only).

GROWTH & CARE: Beginners find this plant difficult, but to experts it is one of the easiest. It needs repotting more often than most plants because it grows so fast. It is one of the few plants with extensive surface roots. For this reason, you should pour the water in 3 or more stages – allowing the first 'pour' to soak in before adding the second pour, etc.

You will find that benjaminas get all 3 pests (mites, mealy bugs and scale), but the plant is durable, so it recovers quickly from a spray. Plants over 8' can be difficult to water (especially in low light). If you have a serious problem with a large tree, report it immediately to your supervisor. A tree can completely defoliate in 10 or 20 days if it is not properly watered. A hot draft (from a heating vent) can defoliate the plant in 2 or 3 days.

RUBBER TREE

Botanical. name: *Ficus elastica 'Decora'*
Trade name: Decora, Rubber
Family & Form: Tree (bushy)
Standard size: 4' to 6' tall
Other sizes: From 1' to 20'+
Minimum light: 50fc (Avg. artificial)
Minimum ideal light: 100fc+ (High artificial)
Care: Very easy, even in low light
Physical durability: Excellent
Resistance to pests: Very good
Similar plants: Croton

WHEN TO WATER: Plants in minimum light (50fc) in a 10" pot.

> BY TOUCH: The soil surface will be dry, but not crumbly.
> MOISTURE METER: Insert 3", and indicator deflects 1/2 or less.
> VISUALLY: Large leaves fold inward as moisture is exhausted.

HOW MUCH TO WATER: In light below 80fc, you moisten just the dry soil (approx. 1 quart every 14 days). In average window light (250fc), a plant in a 10" pot uses approximately 2 quarts every 7 days. Soak the roots thoroughly only when the light is very bright (over 400fc), i.e. a West or South window.

OVERWATERING: Although an extremely durable plant, the Rubber will shed green leaves when overwatered for a month or more in artificial light (below 100fc). Move to higher light if the shedding from overwatering doesn't stop within 3 weeks. In good window light, shedding stops within 3 to 5 days. During acclimatization, the Rubber sheds briefly between the 2nd and 4th weeks.

UNDERWATERING: The Rubber is also difficult to underwater. The leaves will hang more than 45° lower than normal before they start to turn yellow. While the Fig Tree's leaves turn yellow at once, the Rubber's leaves get a mottled yellow before changing to a bright yellow. Soak thoroughly when underwatering is obvious.

GROWTH & CARE: Although all Ficus plants are somewhat difficult for beginners, this one is the easiest. Nursery-grown leaves are always glossy and smooth. In very low light, they become small and winkled. In moderately low light (80fc), they are thin and limp. Like most Ficus, you should pour the water in 2 to 4 stages to allow it to soak straight down into the roots. In bright window light (over 500fc), it uses 10 times as much water as in low light (50fc).

When you prune, be careful or the white sap may drip on the carpet and stain it. The sap also burns the skin so wash up immediately if you get it on your hands. This plant survives temperatures nearly down to freezing and is not bothered by 90°F either. The red-leaved variety 'Rubra' keeps better in low light.

BUDDHIST PINE

Botanical name: *Podocarpus macrophyllus*
Trade name: Podocarpus
Family & Form: Tree (bushy upright)
Standard size: 3' to 6' tall
Other sizes: Occasionally to 12'+
Minimum light: 60fc (Avg. artificial)
Minimum ideal light: 100fc+ (High artificial)
Care: Slightly difficult
Physical durability: Excellent
Resistance to pests: Very good
Similar plants: *Podocarpus nagi*

WHEN TO WATER: Plants in minimum light (60fc) in a 10" pot.

> BY TOUCH: The soil surface will be bone dry before you water.
> MOISTURE METER: Insert halfway down, needle 'pins out'.
> VISUALLY: Needles become limp and hang almost straight down.

HOW MUCH TO WATER: In low light (below 100fc), barely moisten the top half of the soil. Give a 10" plant approx. 1½ quarts every 14 days. This plant needs very little water. Even in bright window light, you will never keep it 'evenly moist'. Feel the needles for limpness before watering.

OVERWATERING: Too much water is extremely serious because many of the needles will turn yellow at the tips. Needles along the stems turn first. Acclimatization (the first 4 weeks) generally sheds green needles. If a plant is getting burned tips, move it immediately to bright light. Trees with soft needles are easier to water than those with coarser needles.

UNDERWATERING: Too little water is also serious, but the reaction is slower than overwatering. Tips of the needles turn brown, and when 25% of the needles have suffered, the plant should be rotated back to the greenhouse. The oldest needles (closest to the stems) are always affected first. Needles of dry plants become brittle and rustle when touched.

GROWTH & CARE: The Podocarpus grows slowly except in bright window light. The main stems are spindly and weak, so you will have to stake them until the plant is 6' or 7' tall. This plant looks poorly when pruned (especially the side branches), thus large plants need a lot of room. Like the Norfolk, it dislikes warm temperatures, although it survives 72°F quite well.

In low light the Podocarpus naturally sheds the needles growing out of the main stems. In bright window light (West and South), it looks almost like a different plant – becoming furry and lush. The originally nursery container is sufficient for 5-10 years' root growth, but after 3 years another inch or two of soil can be added under the rootmass. A 6' plant in a 14" pot will grow 10' tall before it needs a new pot. Feed half as much as a typical Ficus.

NORFOLK PINE

Botanical name: *Araucaria heterophylla*
Trade name: Norfolk
Family & Form: Tree (branching)
Standard size: 3' to 5' tall
Other sizes: From 1' to 16'+
Minimum light: 75fc (Avg. artificial)
Minimum ideal light: 150fc+ (North window)
Care: Difficult in temperatures over 65°F
Physical durability: Fairly good
Resistance to pests: Excellent
Best feature: Very low maintenance

WHEN TO WATER: Plants in minimum light (75fc) in a 10" pot.

BY TOUCH: The soil surface will be dry as far down as you can feel.
MOISTURE METER: Insert 3", don't water if needle deflects any.
VISUALLY: The side branches sag visibly and the needles feel dry.

HOW MUCH TO WATER: You will be surprised at how little water this plant uses. It's always best to pour the water around the trunk, and allow the outer soil to remain dry. Water sparingly, especially below 150fc (approx. 3/4 quart every 14 days). In 350fc, a 10" plant will use approximately 1½ quarts every 7 days.

OVERWATERING: Too much water, even once, in a light below 100fc is deadly. The roots are delicate and even well-grown plants have very few to dry out the soil. Most plants show overwatering in 1 or 2 weeks, but the Norfolk waits 1 or 2 months. By this time, most of the roots will be dead. Lower branches and branchlets fall off from excess water.

UNDERWATERING: Too little water is nearly as fatal as too much. Thus, the Norfolk is definitely one of the most difficult of all tropicals to water. Underwatered plants first develop brown needles along the main trunk. Then the side branches turn brown and crisp. It's not easy to save a plant after more than 10% of the branches have died.

GROWTH & CARE: Most Norfolks, fresh from the nursery, have an underdeveloped root system. A plant in a 10" pot often has a root-to-soil ratio of less than 40%. They abhor warm temperatures, but thrive when the range is 50°F to 60°F. They tolerate cool air-conditioned offices with bright artificial light (100fc+). It takes time to get the knack of maintaining this plant.

The key is slight underwatering. Norfolks grow slowly and you can expect a maximum of one set of new branches every year (they grow a whole new set at once). Plants over 6' tall are much more graceful than the juveniles. It is possible to keep them healthy in temperatures over 70°F, but the humidity must be high and the light must be bright. Develop the habit of feeling the needles for moisture content.

SCHEFFLERA

Botanical name: *Brassaia actinophylla*
Trade name: Shef, Brassaia
Family & Form: Tree (shrubby)
Standard size: 3' to 6' tall
Other sizes: From 1' to 20'+
Minimum light: 50fc (Avg. artificial)
Minimum ideal light: 80fc+ (High artificial)
Care: Fairly easy, especially large plants
Physical durability: Fair (the leaves are brittle)
Resistance to pests: Poor (mites)
Best feature: Large, glossy leaves

WHEN TO WATER: Plants in minimum light (50fc) in a 10" pot.

> BY TOUCH: It will barely feel moist 2" below the soil surface.
> MOISTURE METER: Insert 4", water when indicator is 1/2 or less.
> VISUALLY: The 'hands' of leaves hang 20° below perpendicular.

HOW MUCH TO WATER: For such a large plant, the Shef uses surprisingly little water. In minimum light, water lightly – just around the stems (approx. 1/2 quart every 14 days). In a South or bright West window, you can soak (but not heavily). There, a 10" plant will use approximately 2½ quarts every 7 days.

OVERWATERING: Although the oldest leaves and branches drop to the ground when the plant is overwatered, the shedding stops within 2 weeks when you stop watering. In low light (below 100fc), the plant may not require any water for 3 or 4 weeks. When four or more entire branches fall off, you can tell that the plant is extremely overwatered.

UNDERWATERING: The leaves mottle and turn yellow when the roots are too dry. 50% of them can turn yellow in 3 weeks if the problem is not corrected. Plants fresh from the nursery can disintegrate fast, but older established plants get extremely limp before the foliage starts to suffer. Don't soak an underwatered plant – a light moistening is sufficient.

GROWTH & CARE: 3' to 5' plants, fresh from the nursery, have numerous suckers growing out of the soil. They die back within 8 weeks in low light, so you might as well remove them. Small Shefs are much harder to care for than those over 6' tall, but after 6 months indoors, they get much easier. Spider mites are a serious problem, especially in summer. The leaves turn a mottled yellow (somewhat similar to underwatering), except the mottling occurs over the entire plant – rather than just the lower leaves.

Since the stems snap off easily, be extra careful when cleaning these plants. It's best to keep them away from heavy traffic areas. The Shef grows fairly rapidly, even in artificial light, thus a spindly plant fills in fast. The Tupidanthus looks almost identical, but is 5 times easier to care for since it has better roots.

HAWAIIAN SCHEFFLERA

Botanical name: *Schefflera arboricola*
Trade name: Arboricola
Family & Form: Tree (bushy)
Standard size: 2' to 4' tall
Other sizes: Occasionally to 10'+
Minimum light: 40fc (Low artificial)
Minimum ideal light: 80fc+ (High artificial)
Care: Fairly easy, especially over 120fc
Physical durability: Very good
Resistance to pests: Very good
Best feature: Dense growth

WHEN TO WATER: Plants in minimum light (40fc) in a 10" pot.

> BY TOUCH: The soil surface gets barely dry before you water.
> MOISTURE METER: Insert 2½", when indicator doesn't 'pin out'.
> VISUALLY: The 'hands' of leaves hang 30° below perpendicular.

HOW MUCH TO WATER: This plant is halfway between the Schefflera (small root system) and the Fig Tree (massive root system). Thus, you moisten only the dry soil in light below 100fc (approx. 1 quart every 7 days). Water a 10" plant liberally in bright window light over 300fc+ (approx. 2½ quarts every 7 days).

OVERWATERING: The Hawaiian Shef dries out rapidly, so wet roots cause only a moderate degree of shedding (green leaves). The lowest leaves shed first. Acclimatization (the first 5 weeks) also sheds old leaves. When this plant is overwatered in low light, withhold water for 1 week. In bright light, however, you sprinkle the surface lightly, then resume regular watering the next week.

UNDERWATERING: A few consecutive weeks of underwatering in light below 100fc can result in 10% of the leaves turning yellow. In bright window light (over 400fc), one week of dry roots can mean 30% or more yellow leaves. Since the Hawaiian Shef has numerous surface roots, you should pour the water in 3 or 4 stages to avoid having the water run down the sides of the rootmass.

GROWTH & CARE: This small-leaved plant is one of the easiest for beginners. It has so many leaves that up to 50% of them can be lost through improper watering before its overall appearance is noticeably affected. In fact, the plant often looks better after it sheds (or you remove) the suckering growth and interior leaves.

As the plant grows, the spindly stems are generally too weak to support the masses of leaves that grow from the lengthening branches. As a result, you must either stake the branches or cut them back. Hawaiian Scheffleras need an additional 2" of soil underneath the rootmass every year or two. Pack the soil down around the edges of the pot, or the soil shrinks away from the sides.

FALSE ARALIA

Botanical name: *Dizygotheca elegantissima*
Trade name: Elegantissima
Family & Form: Tree (upright)
Standard size: 3' to 6' tall
Other sizes: From 1' to12'+
Minimum light: 70fc (Avg. artificial)
Minimum ideal light: 250fc+ (East window)
Care: Difficult, especially for beginners
Physical durability: Fair (it sheds)
Resistance to pests: Fair (mites)
Similar plants: The Aralia family

WHEN TO WATER: Plants in minimum light (70fc) in a 10" pot.

> BY TOUCH: Water when the soil surface is crumbly-dry.
> MOISTURE METER: Insert 4", when indicator deflects 1/2 or less.
> VISUALLY: Largest leaves droop a little; newest droop a lot.

HOW MUCH TO WATER: Except in window light over 300fc, this is not an easy plant to water. In artificial light, pour sparingly around the stems – barely moistening the dry soil (approx. 1 quart every 14 days). In a West window (300fc), this plant uses approx. 2 quarts every 7 days.

OVERWATERING: When the light is below 100fc, this plant can shed 3/4 of its leaves in 7 days when overwatered once. Even in a sunny location (over 300fc), it can shed massively if overwatered for 3 straight weeks. When the lowest branches start dropping to the floor, stop watering immediately (for at least 2 weeks), and move the plant to more light.

UNDERWATERING: Roots that are even slightly too dry can cause serious yellow leaves. Except for older plants, the yellowing starts with the smallest leaves. During acclimatization (the first 4 weeks) the plant shows similar symptoms, but this is a busy plant's natural response to less light. An underwatered plant is readily cured by giving it 50% more water than usual.

GROWTH & CARE: The new leaves of this plant start off as a beautiful dark bronze and gradually turn dark green. When the plant reaches 8' tall, the leaves at the top will be 3 times larger than the lower leaves. A plant that has semi-defoliated, but is growing in good light, will sprout new branches. Otherwise, the bare stems will stay that way.

Unfortunately, this beautiful plant is prone to mites and sometimes scale. The mites are hard to manage indoors, because the spray damages the foliage and it's a good policy to return them to the greenhouse for 2-3 months. You can make the False Aralia easier to service by always pouring the water only in the center of the pot. After a few months, the plant grows its new roots directly under the stems and overwatering is less of a problem.

BAMBOO PALM

Botanical name: *Chamaedorea erumpens*
Trade name: Erumpens
Family & Form: Palm (narrow upright)
Standard size: 4' to 6' tall
Other sizes: Up to14'+
Minimum light: 30fc (Low artificial)
Minimum ideal light: 60fc+ (Avg artificial)
Care: One of the easiest Palms
Physical durability: Fairly good
Resistance to pests: Fair (mites)
Similar plants: *Neanthe Bella, Seifrizii*

WHEN TO WATER: Plants in minimum light (30fc) in a 10" pot.

>   BY TOUCH: It will barely feel moist 3" below the soil surface.
>   MOISTURE METER: Insert 4½", when indicator is 1/3 or less.
>   VISUALLY: Water when the fronds droop 45° below horizontal.

HOW MUCH TO WATER: In 30 footcandles, water only the dry soil at the top (approx. 2 cups every 2 or 3 weeks). In 60fc, pour just enough to penetrate to the bottom of the pot. A 10" uses approx. 1 quart a week.

OVERWATERING: Like most Palms fresh from the nursery, an overwatered Bamboo Palm gets yellow tips, but plants that have been indoors over a year get black tips. When yellow tips occur halfway up the length of the plant, it is seriously too wet. Withholding water for 2-3 weeks will stop the burning. This plant uses very little water for its size.

UNDERWATERING: Brown tips mean the roots are slightly too dry, but when some of the lower fronds turn completely yellow, the roots are far too dry. The period of acclimatization (the first 10 weeks), produces yellow fronds (always the lowest fronds first). Up to 20% of the fronds will shed in minimum light (below 60fc), but less than 5% shed in a North window (over 100fc).

GROWTH & CARE: There are several commonly-used Bamboo Palms. The Neanthe Bella is a dwarf, rarely over 6' tall, while the Erumpens and Seifrizii both grow to 14' or larger. The Seifrizii is the more attractive because its frond segments are uniform in size. It also tolerates 25% less light. All 3 plants have sensitive foliage, so feed and clean them cautiously. Overfeeding causes black spots on the leaves and too much plant shine can make the leaves fold under.

Chamaedoreas are exceptional low light plants because they are difficult to overwater. Each plant has 10 or more stems, meaning the pot is generally filled with roots. The Neanthe Bella is the best Bamboo Palm for dimly-lit interiors and will survive 18 to 24 months in 15 footcandles of light. Repotting is rarely necessary. Simply add more soil to the bottom of the pot.

KENTIA PALM

Botanical name: *Howea forsterana*
Trade name: Kentia, Forsterana
Family & Form: Palm (upward curving)
Standard size: 3' to 6' tall
Other sizes: Occasionally to 15'
Minimum light: 50fc (Avg. artificial)
Minimum ideal light: 90fc+ (High artificial)
Care: Average for a Palm
Physical durability: Very good
Resistance to pests: Fairly good
Similar plants: *Howea belmoreana*

WHEN TO WATER: Plants in minimum light (50fc) in a 10" pot.

BY TOUCH: Water when soil is dry as far down as you can reach.
MOISTURE METER: Insert 4", water when indicator deflects 1/2.
VISUALLY: The fronds droop 90° at the tips when it needs water.

HOW MUCH TO WATER: Like most plants with small root systems, low light means you water gingerly around the stems – ignoring the soil at the sides of the pot (approximately 1½ quarts every 2 weeks in 60fc). In good window light (over 400fc), water the entire rootball (a 10" plant uses approx. 2½ quarts every week).

OVERWATERING: Kentia Palms are durable plants and thus withstand 2 or 3 overwaterings without suffering any more than yellow frond tips. The tips turn yellow first, then mushy brown. When a plant is too wet, you may not need to water it again for 4 or 5 weeks. In bright window light, withholding the water for one or two weeks is sufficient.

UNDERWATERING: Underwatering and overwatering are equally serious with this plant. Since the fronds are so heavy, dry soil means the plant can pull its roots out of the soil and topple over. Too little water results in brown tips – or in extreme cases, the whole frond turns yellow. Give a dry plant twice as much water as normal (1 time only).

GROWTH & CARE: In all but a sunny window, the stems of plants fresh from the nursery must be securely staked – otherwise they will lean over between watering and damage some of the roots. Generally, however, his is one of the easiest plants to care for. It resists pests and survives low light by virtue of its combination of tough, leathery fronds and large strong roots.

Like most Palms, the smaller stems die back in light under 100fc. They will shed the roots growing close to the soil surface because you must dry the soil halfway down in low light. After the first 3 months indoors, the Kentia gradually become symmetrical, producing 4 or 5 new fronds each year. The older fronds start to mottle eventually, and a healthy plant rarely has more than 10 good fronds at a time.

ARECA PALM

Botanical name: *Chrysalidocarpus lutescens*
Trade name: Areca
Family & Form: Palm (upward curving)
Standard size: 4' to 6' tall
Other sizes: From 1' to 15'+
Minimum light: 70fc (Avg. artificial)
Minimum ideal light: 160fc+ (East window)
Care: A poor commercial plant
Physical durability: Fair (it burns)
Resistance to pests: Poor (mites)
Best feature: Low price

WHEN TO WATER: Plants in minimum light (70fc) in a 10" pot.

> BY TOUCH: Water when soil is dry as far down as you can reach.
> MOISTURE METER: Insert 6", water when indicator deflects 1/2.
> VISUALLY: The fronds wilt significantly and the tips droop down.

HOW MUCH TO WATER: The Areca survives fairly low light considering its grassy leaves. Pour the water sparingly around the stems (approx. 3/4 quart every 14 days in 80fc), but water liberally in a window. In 400fc, dry only the top 1" of soil before watering (approx. 2½ quarts every 7-10 days.

OVERWATERING: When the roots are too wet, the tips of the fronds turn yellow, starting with the fronds nearest the pot. Even when you cut back on the water immediately, the burning continues for two or three more weeks. When a plant is acclimatizing (the first 2 months), the small suckers that sprout from the soil also turn yellow.

UNDERWATERING: Dry roots result in brown frond tips. Extreme underwatering will cause an entire frond to turn yellow. When this happens to one of the 3 or 4 largest fronds, the roots are far too dry. Soak the plant immediately and the symptoms will stop within 2 weeks. In low light, the small suckers always die because their roots are at the top.

GROWTH & CARE: The Areca Palm is not a good commercial plant (except in certain specific situations), even though it sells well at the consumer level. It's a poor plant because it needs high humidity to keep from burning at the tips, and because smaller plants (under 7' tall) will shed 70% of their stems in all but well-lit indoor spaces.

The best solution for dealing with the standard 5' Areca – fresh from the nursery – is to cut away all the small sprouts under 2' tall. The plant will be less bushy initially, but the 3 or 4 large stems remaining will eventually grow into a respectable plant. Plants over 7' tall are usually composed of 2 or 3 large stems. These large stems are more mature and suffer less from low humidity. They are much easier to water.

FISHTAIL PALM

Botanical name: *Caryota mitis*
Trade name: Fishtail, Caryota
Family & Form: Palm (upward curving)
Standard size: 5' to 7' tall
Other sizes: Up to 16'+
Minimum light: 60fc (Avg. artificial)
Ideal light: 275fc+ (West window)
Care: Easy in good light
Physical durability: Very good
Resistance to pests: Fair (mites)

DESCRIPTION: These sturdy plants are often wider than they are tall, so their use is limited indoors. Although they tolerate low light, the tips of the fronds burn (yellow) from lack of humidity. They can dry out halfway down between watering. A plant in a 14" pot requires approx. 3 qts. every 14 days in 100fc. Physically, the Fishtail and Fan Palms are the sturdiest indoor plants.

FAN PALM

Botanical name: *Livistona chinensis*
Trade name: Livistona
Family & Form: Palm (downward curving)
Standard size: 2½ ' to 7' tall
Other sizes: Up to 20'+
Minimum light: 60fc (Avg. artificial)
Ideal light: 150fc+ (East window)
Care: Very easy in all light
Physical durability: Excellent
Resistance to pests: Fairly good

DESCRIPTION: Tough, leathery leaves make this plant exceptionally valuable indoors. It is seldom used because it is coarse-looking as a smaller plant. It grows slowly, even in good light. Keep it very dry and it will co-operate by surviving for years in low light. It requires approx. 2½ quarts of water every 21 days in 100fc.

PYGMY DATE PALM

Botanical name: *Phoenix roebelenii*
Trade name: Roebelenii
Family & Form: Palm (downward curving)
Standard size: 2½ ' to 4' tall
Other sizes: Occasionally to 12'
Minimum light: 80fc (High artificial)
Ideal light: 150fc+ (East window)
Care: Fairly easy in good light
Physical durability: Fairly good
Resistance to pests: Poor (mites)

DESCRIPTION: This is a very beautiful and graceful Palm, whose chief disadvantage is its grasslike fronds. The plant is sturdiest when over 6' tall and can single-handedly decorate a large open space. The Roebelenii is even worse than the Areca or Schefflera for attracting mites. A plant in a 14" pot requires approximately 1½ quarts of water every 7 days in 100fc.

SAGO PALM

Botanical name: *Cycas revoluta*
Trade name: Cycas, Sago
Family & Form: Palm (downward curving)
Standard size: 1½' to 3' tall
Other sizes: Rarely over 6-8'
Minimum light: 60fc (Avg. artificial)
Ideal light: 150fc+ (East window)
Care: Easy when kept dry
Physical durability: Excellent
Resistance to pests: Excellent

DESCRIPTION: This plant's leaves are so thick they feel like plastic. Smaller table-sized plants have exceptional keeping qualities. The Sago may produce a set of new growth every year in good light. The key is to completely dry out the soil before watering – like a Cactus. This plant may get by on 2 quarts every 60 days in 100fc.

YUCCA

Botanical name: *Yucca elephantipes*
Trade name: Yucca
Family & Form: Cane (upright)
Standard size: 3½ ' to 6' tall
Other sizes: Often to 14'+
Minimum light: 50fc (Avg. artificial)
Minimum ideal light: 100fc+ (High artificial)
Care: Very easy if kept dry
Physical durability: Excellent
Resistance to pests: Fairly good
Similar plants: Other Yucca species

WHEN TO WATER: Plants in minimum light (50fc) in a 10" pot.

BY TOUCH: Water when soil is dry as far down as you can reach.
MOISTURE METER: Insert 5", water when indicator doesn't move.
VISUALLY: Then outer 2" of the leaves turn downward.

HOW MUCH TO WATER: Although the Yucca prefers bright light, it tolerates artificial light when it is kept very dry (like the cane-type Dracaenas). Pour the water carefully around the stems, barely moistening the dry soil (approx. 1½ quarts every three weeks in 80fc of light).

OVERWATERING: Since this plant reacts slowly to overwatering, the leaves will not show yellow tips for at least six weeks. Sometimes the tips turn black instead of yellow. An obviously overwatered plant will not need water for at least 6-8 weeks. It may not recover when the light is less than 60fc. In bright light, it recovers in two months.

UNDERWATERING: The tips of the leaves turn brown (or the whole leaf turns yellow) when the roots are too dry. However, the leaves will look extremely wilted before this happens. Simply give an underwatered plant a healthy dose of water and it will recover in 2-4 weeks. In bright window light (over 500fc), you have to water thoroughly – twice in a row.

GROWTH & CARE: Most Yuccas are grown from cane sections like the cane-type Draecenas. Occasionally you may find the bush-form. Since nursery-fresh cane-types have the same lack of roots as the Dracaenas, you must be careful never to allow the roots to stay overly wet – especially in low light – or the stem may get mushy and the plant will die.

These plants are especially resistant to pests; but occasionally you may find a grayish moldy growth on the leaves. Spray with a copper-based solution. Generally, you will notice that mites or scale hardly affect a Yucca's growth – whereas most plants suffer serious damage immediately. In bright light a Yucca's leaves are stiffer, and their serrated edges are sharp enough to cut bare skin and snag loose clothing.

CORN PLANTS (Bush-type)

Botanical name: *Dracaena fragrans massangeana*
Trade name: Fragrans or Massangeana
Family & Form: Cane (clump)
Standard size: 4 ' to 7' tall
Other sizes: From 18" to 20'+
Minimum light: 25-30fc (Low artificial)
Minimum ideal light: 40-60fc+ (Avg. artificial)
Care: Easy when watered correctly
Physical durability: Good to Excellent
Resistance to pests: Very good
Similar plants: *Dracaenas 'Warneckei', 'Janet Craig'*

WHEN TO WATER: Plants in minimum light (30fc) in a 10" pot.

> BY TOUCH: Water when soil is dry as far down as you can reach.
> MOISTURE METER: Insert 7", water when indicator is 1/3 or less.
> VISUALLY: The leaves develop deep scallops along the edges.

HOW MUCH TO WATER: Corn Plants have extremely tough leaves and roots – so they use relatively little water. However, when you do water, you must moisten the entire rootmass thoroughly (approx. 3 quarts every 14 days in 40fc). In a window approx. 4 quarts every 7 days.

OVERWATERING: Like Palms, overwatering shows up as yellow leaf tips. If the leaves burn for more than 1/5 of their length, the plant is much too wet. Plants fresh from the nursery always burn, but only the lower 20% of the leaves. The top 25% of the leaves on a plant should never turn yellow or the plant is so overwatered it may fail.

UNDERWATERING: Again, like the Palms, dry roots show up as brown leaf tips. When extremely dry, the lower leaves completely turn yellow. In the nursery, these plants have a flat leaf surface, but indoors the reduced moisture give the leaves a ruffled edge. Soak a dried-out plant thorough (one time only), then resume regular watering.

GROWTH & CARE: There are three major Dracaenas with broad, strap-like leaves. The Massangeana has the widest and longest leaves – striped lengthwise in lime green. The Janet Craig has glossy, dark green leaves about 2/3 the size of the Massangeana's leaves. The Warneckei's leaves have silver gray stripes and they are the smallest of the three Dracaena's leaves.

These three plants are the Interiorscaper's ace-in-the-hole for low light care. When damaged by over or underwatering, they recover quickly. The Massangeana is the most durable, while the other two run a close second. The Warneckei has a problem with brittle leaves, while the Janet Craig's leaves mark up easily from a number of problems. Always keep an eye on the new growth. It should never be pale or sickly or you have overwatered.

CORN PLANTS (Cane-type)

Botanical name: *Dracaena fragrans massangeana*
Trade name: Fragrans or Massangeana
Family & Form: Cane (upright)
Standard size: 3½ ' to 8' tall
Other sizes: Often to 16'+
Minimum light: 30-40fc (Low artificial)
Minimum ideal light: 45-55fc+ (Avg. artificial)
Care: Slightly more difficult than Bush-type
Physical durability: Very good
Resistance to pests: Very good
Similar plants: *Dracaenas 'Warneckei', 'Janet Craig'*

WHEN TO WATER: Plants in minimum light (35fc) in a 10" pot.

    BY TOUCH: Water when soil is dry as far down as you can reach.
    MOISTURE METER: Insert 5", water when indicator is 1/3 or less.
    VISUALLY: The leaves develop deep valleys along the edges.

HOW MUCH TO WATER: Never soak these plants even in bright window light. They must be kept fairly dry because they have relatively few leaves. In low light, pour the water slowly and carefully just around the canes (approx. 1½ quarts every 14 days in 50fc).

OVERWATERING: The cane-type Corn Plants are much easier to overwater than their bush-type brothers. You can wiggle the stems to see how many roots the plants have. The leaves get yellow tips from too much water. When the new growth gets mushy light-brown tips, the plant is much too wet and you must withhold water for at least 4-6 weeks.

UNDERWATERING: Brown leaf tips mean dry roots and when several leaves turn entirely yellow the roots are far too dry. If you look closely, you will notice that each 'head' grows five or more dwarfed leaves before growing the full-sized leaves. The plant sheds these leaves gradually (fastest in low light) and you need not be concerned.

GROWTH & CARE: Usually just the Massangeana is produced as a cane-type plant but occasionally you will see a Janet Craig or a Warneckei in this form. Canes are grown by cutting the stems of tall plants into sections. These sections are stuck in a pot and in about six months the tops of the canes sprout leaves while the bottoms of the canes sprout roots.

This method of nursery production causes the plants to have an undersupply of roots for the first year. Thus they need less water and generally more light. The key to watering is to pour around the canes and let the soil around the pot stay dry. This technique enables the dry soil to pull moisture away from the center of the pot, making overwatering more difficult (especially in low light).

MARGINATA

Botanical name: *Dracaena marginata*
Trade name: Marginata
Family & Form: Cane (upright)
Standard size: 3' to 5½ ' tall
Other sizes: Often to 20'+
Minimum light: 30fc (Low artificial)
Minimum ideal light: 55fc+ (Avg. artificial)
Care: Easy to overwater
Physical durability: Fairly good
Resistance to pests: Fairly good
Similar plants: *Dracaena reflexa*

WHEN TO WATER: Plants in minimum light (30fc) in a 10" pot.

>BY TOUCH: Water when soil is dry as far down as you can reach.
>MOISTURE METER: Insert 5", water when indicator is 1/2 or less.
>VISUALLY: Outer 1/4 of the leaves droop downward noticeably.

HOW MUCH TO WATER: In light below 100fc, you must pour the water lightly around the stems (approx. 1 quart every 14 days). Water just the soil that has dried since the last watering. In bright window light (over 400fc), water the entire rootmass, but do not soak (approximately 1¾ quarts every seven days).

OVERWATERING: This plant is much easier to overwater than the large bush-type Dracaenas because the leaves are thin and delicate. The tips turn yellow, often for 1/3 of their length when the roots are just slightly too wet. Extreme overwatering will cause the stems to get soft and mushy (takes 3 or 4 months), and then the plant is a write-off.

UNDERWATERING: Brown tips show that the roots are too dry. When the roots have been too dry for 14 to 21 days, up to 15% of the lower leaves may turn yellow. Fortunately, one thorough watering cures the problem immediately. In low light, be careful not to soak an underwatered plant or it will show overwatering symptoms within 10-14 days.

GROWTH & CARE: The Marginata is a unique plant. Fresh from the nursery, the leaves are stiff and point slightly upward. Gradually they begin to angle downward. The new leaves (except in extremely bright window light) will always arch gracefully downward from the stem to the tip. You can expect this plant to shed all its nursery-grown leaves within the first year. The 'new' plant is always more attractive.

You are well-advised to keep this plant on the dry side. When the lower leaves of a nursery-fresh plant start to burn, remove them rather than trimming them. They will always continue to burn. The 'heads' of this plant are the most brittle of all tropicals and you will probably snap a few off before you learn to be careful. Fortunately, the broken stem will sprout 1-5 new heads.

CHINESE EVERGREEN

Botanical name: *Aglaonema species*
Trade name: Aglaonema or species name
Family & Form: Cane (bushy)
Standard size: 2' to 3½ ' tall
Other sizes: Rarely to 6'
Minimum light: 35fc (Low artificial)
Minimum ideal light: 60fc+ (Avg. artificial)
Care: Easy when kept dry
Physical durability: Very good
Resistance to pests: Fairly good
Similar plants: Dieffenbachias

WHEN TO WATER: Plants in minimum light (35fc) in a 10" pot.

> BY TOUCH: Water when soil is dry as far down as you can reach.
> MOISTURE METER: Insert 5", water when indicator is 1/3 or less.
> VISUALLY: The outer 3" of each leaf droops downward noticeably.

HOW MUCH TO WATER: Like the rest of the Canes, dryness is preferable to even small degrees of overwatering. Water thoroughly only in extreme bright window light. In very low light, pour the water in the center of the pot around the stems (approx. 1½ quarts every 21 days).

OVERWATERING: The Aglaonema's leaves are semi-succulent, so they turn mushy yellow at the tips from too much water. Extreme overwatering causes the main stems to rot. When the soil refuses to dry out for 3 or 4 weeks, move the plant next to a window or it will fail quickly.

UNDERWATERING: While most Canes and Palms get brown leaf tips when the roots are too dry, this plant is apt to show two or three completely yellow leaves. Usually, however, they will wilt enough so you can see there is a problem. Be careful not to heavily water a dried-out plant because it uses water slowly and may be easily overwatered.

GROWTH & CARE: This plant looks nearly identical to the Dieffenbachia but the Dief is only half as sturdy – especially in artificial light. The Aglaonema's chief drawback is size. It is rarely produced over 4' tall, whereas the Dieffenbachia grows to any size. It also grows at half the speed of the Dieffenbachia so it takes years to develop a sizable plant.

You will notice that an Aglaonema is actually 10 or more different-sized plants in one pot. In low light (just like the Palms), the smaller plants die back. You can remove them before they die back – and still be left with a healthy, attractive plant. This plant gets mealy bugs more often than other pests, but because it has relatively leathery leaves, a single heavy spray will kill the bugs without harming the plant.

SPIDER PLANT

Botanical name: *Chlorophytum comosum*
Trade name: Chlorophytum, Spider
Family & Form: Vine (trailing)
Standard size: 10" hanging basket
Other sizes: 4" to 12" basket
Minimum light: 60fc (Avg. artificial)
Minimum ideal light: 100fc+ (High artificial)
Care: Easy
Physical durability: Fairly good
Resistance to pests: Fairly good
Similar plants: Small rosette-type tropicals

WHEN TO WATER: Plants in minimum light (60fc) in a 10" pot.

BY TOUCH: Water when soil is dry as far down as you can reach.
MOISTURE METER: Insert 3", water when indicator is 1/4 or less.
VISUALLY: When the leaves are noticeably softer and limper.

HOW MUCH TO WATER: These plants have fleshy roots that store water, so you can allow them to get bone dry in low light without damaging the plant. In bright window light, they still must get dry 1/4 of the way down before you water. In 60fc, a 10" requires approximately 2½ cups of water every 14 days.

OVERWATERING: The tips of the oldest (largest) leaves get yellow. Pull off the whole leaf rather than trim. When 20%+ of the leaves have yellow tips, the problem is serious. Withhold water for 3 or 4 weeks to correct the problem. Wait one week past the time the burning stops before watering. This plant can survive 8 weeks without water.

UNDERWATERING: Spiders get brown leaf tips when the roots are too dry and because of low humidity. You can expect the oldest leaves to brown on a regular basis. Be concerned only when more than 10% of the tips are turning brown. Then you know for sure the plant is underwatered. Brown tips are a fact of life with this plant.

GROWTH & CARE: Spider plants are easy to maintain when you remember to keep them very dry. In the greenhouse they grow large coarse leaves but indoors the leaves are smaller and more delicate. Therefore, as new leaves grow, you can remove the oldest ones at the same rate. This plant can grow a new set of leaves in less than one year.

Cosmetically, you will have to decide whether the plant is more attractive with or without the runners. If you decide to keep them, the plant people in the office will be happy when you occasionally cut off a few large ones for them to propagate. Cuttings will grow in a 4" pot of soil but tell people to withhold water for at least 2-4 weeks. This will give the small roots a chance to get accustomed to their new environment. Otherwise, they may rot.

SPLIT LEAF

Botanical name: *Monstera deliciosa*
Trade name: Monstera, Split-Leaf
Family & Form: Vine (upright)
Standard size: 2½' to 4½'
Other sizes: Occasionally 6' to 8'
Minimum light: 40fc (Low artificial)
Minimum ideal light: 80fc+ (High artificial)
Care: Reasonably easy for beginners
Physical durability: Fairly good
Resistance to pests: Fairly good
Similar plants: Large-leaved Philodendrons

WHEN TO WATER: Plants in minimum light (40fc) in a 10" pot.

> BY TOUCH: Water when soil is dry as far down as you can reach.
> MOISTURE METER: Insert 6", water when indicator is 1/2 or less.
> VISUALLY: Water when the leaves wilt noticeably.

HOW MUCH TO WATER: Most Philodendrons prefer to have their roots dry out considerably, even in bright light. They are easy to overwater because they have so few roots in relation to their mass of foliage. In low light, pour the water only around the stems (approximately 1½ quarts every 14 days in 40fc).

OVERWATERING: The edges of leaves turn yellow, then black, when the roots are too wet. In lower light levels, the oldest leaves turn yellow then black, so you need only be concerned when the uppermost leaves are suffering. It may take 4 or 5 weeks for an overwatered plant to dry out. If it doesn't dry within 6 weeks, the plant is finished.

UNDERWATERING: Since the leaves get soft and wilted before they turn completely yellow, you will rarely underwater this plant. Even when it is much too dry, pour only 50% more water than normal as the plant can go from underwatered to overwatered quickly. The leaves on the smallest stems wilt first because their roots are close to the soil surface.

GROWTH & CARE: The Split-Leaf needs to be staked for the leaves to split. All Philodendrons grow larger leaves when the stems are securely supported. In nature, these plants cling to other plants as they grow, allowing the stems to grow aerial roots. You can cut off these aerial roots if the plant is more attractive without them, however, they do help to gather additional nutrients.

Philodendrons rarely ever need repotting, but if you add any soil, try using bark. This plant grows best in bar. In low light, the new growth is apt to be pale and sickly. Cut it off unless the leaves are needed to enhance the overall appearance. The Split-Leaf is an ideal plant for hot, dry conditions. It thrives where lack of humidity destroys other tropicals.

POTHOS

Botanical name: *Epipremnum aureum*
Trade name: Pothos, Epipremnum
Family & Form: Vine (trailing)
Standard size: 10" hanging basket
Other sizes: 4" to 12" basket
Minimum light: 25fc (Low artificial)
Minimum ideal light: 60fc+ (Avg. artificial)
Care: Moderately easy
Physical durability: Fairly good
Resistance to pests: Very good
Similar plants: Small-leaved Philodendrons

WHEN TO WATER: Plants in minimum light (25fc) in a 10" pot.

    BY TOUCH: Water when soil is dry as far down as you can feel.

    MOISTURE METER: Insert 3", water when indicator is 1/2 or less.

    VISUALLY: Water when the leaves get limp and fold inward.

HOW MUCH TO WATER: In bright window light, these are easy plants to water. You never soak them but they respond well to a light-handed dousing. In low light (30fc), however, you have to pour the water in dribbles just around the stems (approximately 1/2 cup every 7 days, or 3/4 cup every 14 days).

OVERWATERING: The leaves closest to the pot will get mushy and then turn black when the roots are too wet. When this plant is kept overly moist for more than 3 or 4 weeks, there is a danger that the roots will rot quickly and the stems will put out of the soil. If this plant refuses to dry out in 3 or 4 weeks, move it to more light or it will die in short order.

UNDERWATERING: The leaves closest to the stem wilt, then turn yellow, and finally turn brown when the roots are too dry. The symptoms of overwatering and underwatering are somewhat similar initially, so you have to watch closely to tell the difference. Don't soak a dry plant – give it 50% more water than normal and it will recover in 12 hours.

GROWTH & CARE: Although tropical plants respond best when they are watered at the "point of wilting", the Pothos, with its small root system, often needs no more than a small dribble of water every week in low light. This is because the pot is small and only rarely can a plant go 2 weeks without water. When you check and find there is still sufficient moisture in the soil, pour a little water around the stems in any case. Otherwise, it is sure to be wilted the next time you visit.

A Pothos that has been growing in an interior space for more than a year is much weaker than it was when first installed. You will find that a plant in 50fc, with 40 or more leaves, may need only 2 tablespoons of water every week. The Cordatum, which looks similar but has solid green leaves, uses even less water.

NEPHTHYTIS

Botanical name: *Syngonium podophyllum*
Trade name: Syngonium , Nephthytis
Family & Form: Vine (trailing)
Standard size: 10" hanging basket
Other sizes: 4" to 12" basket
Minimum light: 35fc (Low artificial)
Minimum ideal light: 60fc+ (Avg. artificial)
Care: Very easy for a vine
Physical durability: Fairly good
Resistance to pests: Fairly good
Similar plants: Other similar-leaved Vines

WHEN TO WATER: Plants in minimum light (35fc) in a 10" pot.

BY TOUCH: Allow soil to dry completely before watering.
MOISTURE METER: Insert 3", water when indicator is 1/3 or less.
VISUALLY: All the leaves will hang limply when it needs water.

HOW MUCH TO WATER: This excellent low light plant can go longer without water than any other Vine. In low light, dribble water carefully around the stems. In a sunny window it will use 10 times as much water. It requires approximately 2 cups of water every 14 days in 50fc of light.

OVERWATERING: Look for this plant to develop yellow leaf margins from roots that have been wet for too long. It may take 3 or 4 weeks for the symptoms to show up, so cut back on the water as soon as you see signs. In low light, it is normal for the leaves closest to the pot to die back gradually. If they show only slight overwatering, ignore the signs.

UNDERWATERING: Like all large-leaved Vines, dry roots mean limp leaves, but the leaves will become quite soft before they turn completely yellow. This plant perks up in 6-8 hours after watering. Be careful never to soak in light below 100fc, or an underwatered plant quickly becomes too wet. If one of the stems is wilted, and the others are not, water carefully around the affected stem.

GROWTH & CARE: This plant's major appearance problem in low light is the large distance between each successive leaf. When the distance is greater than 6", cut the stem back to a more attractive length. The leaves also get quite small in extremely low light, so you may cut them right back to the soil level in order to promote uniform new growth.

Be careful when the Nephthytis is hanging close to an air-conditioning duct, because the large leaves will dry out faster than the roots can pump moisture. This plant will actually tolerate less light than the minimum given above, but the growth becomes so unattractive, the result is not worth the effort. As with all of the low light Vines, enhance the plant's performance by putting it in a clay pot in light levels below 60fc.

GRAPE IVY

Botanical name: *Cissus rhombifolia*
Trade name: Cissus or species name
Family & Form: Vine (trailing)
Standard size: 10" hanging basket
Other sizes: 4" to 12" basket
Minimum light: 30fc (Low artificial)
Minimum ideal light: 50fc+ (Avg. artificial)
Care: Moderately difficult
Physical durability: Fair (delicate)
Resistance to pests: Fairly good
Similar plants: Other 'woody' Vines

WHEN TO WATER: Plants in minimum light (30fc) in a 10" pot.

>BY TOUCH: Allow soil to dry down at least 2" before watering.
>MOISTURE METER: Insert 5", water when indicator is 1/3 or less.
>VISUALLY: All the leaves will be limp when it needs water.

HOW MUCH TO WATER: Like all Vines that tolerate low light, the Grape Ivy must be watered by dribbling the water carefully around the stems. Otherwise, it is easy overwatered. Never soak, even in front of a sunny window. A 10" plant uses approximately 1 cup of water every 14 days in 40fc of light.

OVERWATERING: When the roots of nursery-fresh plants are too wet, the leaves turn black. When the roots of well-established plants are too wet, the leaves turn a mottled yellow. Be careful not to overwater because these plants can be damaged even faster than the Pothos. If a Grape Ivy refuses to dry out in 3 or 4 weeks, move it into more light.

UNDERWATERING: The leaves normally get brown and crisp from too little water, but fresh from the nursery, they turn bright yellow before turning brown. Since this plant has delicate roots when young, give it only 25% more water than usual when you discover the roots are overly dry. In low light, this plant is nearly impossible to underwater.

GROWTH & CARE: Since the Grape Ivy has delicate, fragile leaves, some of them always turn brown when it is first installed in an account. You will notice that the clusters of 3 leaves always die at the same time. Also, some of the individual stems will die back occasionally. You can tell a stem is dying when it starts to turn a darker brown. Even though this plant is fragile, it grows beautifully after it adjusts to a new location. The plant grows wild in bright light, or prune it regularly so that it turns into a leafy, compact ball. Cut it back in low light or the runners become gangly. After these plants have been indoors for 18 months or more, their stems thicken and get stronger and you will have much less trouble with them because they are more difficult to overwater.

CREEPING FIG

Botanical name: *Ficus pumila*
Trade name: Pumila, repens
Family & Form: Vine (trailing)
Standard size: 10" hanging basket
Other sizes: 4" to 12" basket
Minimum light: 80fc (High artificial)
Minimum ideal light: 150fc+ (North window)
Care: Easy if it doesn't dry out
Physical durability: Very good
Resistance to pests: Fairly good
Similar plants: Other 'woody' Vines

WHEN TO WATER: Plants in minimum light (80fc) in a 10" pot.

>BY TOUCH: The soil surface will be barely moist before watering.
>MOISTURE METER: Insert 1", water when indicator is 1/2 or less.
>VISUALLY: The leaves get slightly paler when it needs water.

HOW MUCH TO WATER: Water the Creeping Fig exactly like you water the Fig Tree. In low light, moisten only the soil that has dried since your last visit. In a sunny window, keep moist but not soaking wet. A 10" plant requires approx. 3 cups of water every 7 days in 100fc.

OVERWATERING: Although not easy to overwater, the Creeping Fig gets black leaves when the roots are too wet. When this happens, give the plants half as much water as you normally would, and the problem will be corrected in one week. The leaves of older plants often show overwatering when the outer half of the leaves turn brown.

UNDERWATERING: You better hope you never underwater this plant, because it can become totally crisp and brown in 2 or 3 days after drying out. The first signs are shriveled green leaves closest to the soil. Then the whole plant may suddenly turn bright yellow, and finally all the leaves will dry up. If they haven't all dried up, a good soak corrects the problem in 4-6 hours.

GROWTH & CARE: With proper care, this is a sturdy, fast-growing Vine. It easily climbs up walls, twines around a pillar, or blankets the soil in a garden. Because it grows so fast, it benefits from extra fertilizer, especially in bright light. Under the right conditions, the runners can grow more than 12 feet a year.

Surprisingly, these plants are slow to outgrow their pot, and you can easily prolong putting them in a bigger pot by simply adding 1½" of fresh soil underneath the rootball. Make sure you loosen the roots at the bottom and sides to encourage faster migration into the new soil. English Ivy (*Hedera helix*) reacts nearly identically to the Creeping Fig. They take a little less water and prefer slightly higher humidity.

PRAYER PLANT

Botanical name: *Maranta leuconeura*
Trade name: Maranta
Family & Form: Vine (trailing)
Standard size: 10" hanging basket
Other sizes: 4" to 12" basket
Minimum light: 60fc (Avg. artificial)
Minimum ideal light: 100fc+ (High artificial)
Care: Moderately difficult
Physical durability: Fairly good
Resistance to pests: Fairly good
Similar plants: Calathea species

WHEN TO WATER: Plants in minimum light (60fc) in a 10" pot.

> BY TOUCH: Allow the upper soil to get bone dry before watering.
> MOISTURE METER: Insert 3", water when indicator is 1/3 or less.
> VISUALLY: The leaves droop noticeably when it needs water.

HOW MUCH TO WATER: In average artificial light, you must dribble the water around the stems to avoid overwatering. Never soak, even in a sunny window. In spite of the size of the leaves, this plant uses very little water. A 10" plant uses approximately 2 cups of water every 14 days in 100fc of light.

OVERWATERING: The margins of the leaves turn yellow when the roots are too wet. A plant fresh from the nursery always gets yellow leaf margins (on the oldest leaves), except when the light is very bright (over 250fc). You may have to withhold water 3 or 4 weeks when the roots are overly moist. Constant overwatering can destroy the roots in 6-8 weeks.

UNDERWATERING: The leaf margins turn brown, both from underwatering and low humidity. The oldest leaves always turn color first. Extreme underwatering causes whole leaves to turn yellow. When a plant has 6 or more yellow leaves, the roots are very dry. Give it a maximum of 25% more water than normal in this situation because these plants are easily overwatered.

GROWTH & CARE: This is not an easy plant for beginners to maintain because there is a narrow range between its minimum and maximum water requirements. Most tropical plants can be miswatered by 10-20% before the leaves are damaged. The margin for error on a Prayer Plant is 5% or less. It prefers very high humidity or the leaf margins burn with disconcerting regularity.

New growth is much smaller and less vividly colored than the nursery-grown leaves. Its best advantage is the small amounts of water it requires. When you have mastered the care of a Prayer Plant in artificial light, you can consider yourself an Interiorscape expert.

LIPSTICK PLANT

Botanical name: *Aeschynanthus lobbianus*
Trade name: Aeschynanthus
Family & Form: Vine (trailing)
Standard size: 10" hanging basket
Other sizes: 6" to 12" basket
Minimum light: 80fc (High artificial)
Min. ideal: 100fc+ (North window)
Care: Moderately easy
Physical durability: Fairly good
Resistance to pests: Fairly good
Similar plants: Other fleshy Vines

DESCRIPTION: The Lipstick is an exceptional flowering plant. It grows in relatively low light, for a Vine, and it usually flowers twice a year when it grows next to a window. It's easy to water because the fleshy leaves store water and get noticeably flabby between watering. This plant requires approx. 2 cups of water every 14 days in 100fc of light.

WAX PLANT

Botanical name: *Hoya carnosa*
Trade name: Hoya or species name
Family & Form: Vine (trailing)
Standard size: 8" hanging basket
Other sizes: Occasionally 10" basket
Minimum light: 50fc (Avg. artificial)
Min. ideal light: 100fc+ (High artificial)
Care: Easy when kept dry
Physical durability: Very good
Resistance to pests: Excellent
Similar plants: All Hoya species

DESCRIPTION: Regrettably, Hoyas are difficult to find – mostly because they grow slowly, making them expensive to produce. As hanging plants, they are hard to beat because they take very little water and never shed. Water them like a Succulent, by feeling the leaves for moisture. When the leaves are soft, they need water (approx. every 14-21 days in 100fc).

SWEDISH IVY

Botanical name: *Plectranthus australis*
Trade name: Plectranthus
Family & Form: Vine (trailing)
Standard size: 10" hanging basket
Other sizes: 6" to 12" pot
Minimum light: 80fc (High artificial)
Min. ideal light: 200fc+ (North window)
Care: Moderately easy
Physical durability: Fair (weak stems)
Resistance to pests: Very good
Similar plants: Creeping Charlie

DESCRIPTION: The Swedish Ivy is the best of the semi-succulent Vines for lower light levels. It continually sprouts new growth close to the pot – making it an easy Vine to control. Like all succulent-types, feel the leaves before watering. All Succulents will rot quickly when the roots are too wet. It uses approx. 1½ cups of water every 14 days in 100fc.

INCH PLANT

Botanical name: *Tradescantia species*
Trade name: Tradescantia
Family & Form: Vine (trailing)
Standard size: 10" hanging basket
Other sizes: 8" to 12" pot
Minimum light: 100fc (High art.)
Min. ideal: 250fc+ (East window)
Care: Difficult to keep attractive
Physical durability: Fair (weak stems)
Resistance to pests: Fairly good
Similar plants: Zebrina species

DESCRIPTION: The Inch Plant is popular as a houseplant because it is inexpensive and the leaves are very colorful. Commercially it is a disaster, except when used as a groundcover in a well-lit planter garden, where it is easily managed and gives fast coverage. It uses less water than you might expect (approx. 3 cups every week in 250fc).

BOSTON FERNS

Botanical name: *Nephrolepis exaltata 'Bostoniensis'*
Trade name: Species name
Family & Form: Fern (trailing)
Standard size: 10" hanging basket
Other sizes: 6" to 12" basket
Minimum light: 70fc (Avg. artificial)
Minimum ideal light: 100fc+ (High artificial)
Care: Easiest when kept moist
Physical durability: Fair (brittle)
Resistance to pests: Very good
Similar plants: All Boston Fern types

WHEN TO WATER: Plants in minimum light (70fc) in a 10" pot.

    BY TOUCH: Keep the soil surface moist at all times.
    MOISTURE METER: Insert 1", water unless indicator 'pins out'.
    VISUALLY: Fronds get noticeably lighter in color as plant dries.

HOW MUCH TO WATER: In a sunny window, it's impossible to overwater a healthy Boston Fern. In 80fc, you must let them dry out slightly (approx. 3 cups of water every 14 days). Water the entire soil surface evenly, like you would the Fig Tree. When the light is more than 400fc, Bostons need water twice a week.

OVERWATERING: Even though it's difficult to give a healthy Fern too much water, they will suffer after 4-6 weeks of soggy roots. The symptoms are blackened, broken fronds, or sometimes the fronds become a pale sickly green. The remedy is to water sparingly for the next month or two, and give the plant a chance to grow new roots.

UNDERWATERING: Many Ferns are underwatered because the water is not applied slowly enough. The first stage of underwatering is faded green leaves. After this, the fronds break and turn brown. The solution is a good soak, but the plant will take 4 to 6 months to grow back. Cut off (at the base) all the fronds with broken stems, because they will die eventually, even if they are still green.

GROWTH & CARE: Ferns can be troublesome for beginners. But, the experienced Interiorscaper finds them one of the easiest tropicals – because they grow fast and thus can be shaped and controlled without difficulty. To be successful with a Fern, you must learn patience in watering. The soil surface is always tightly-packed with roots and the water has a tendency to run down the sides of the pot, rather than penetrating into the center of the rootmass.

    A Fern's fronds are quite fragile, so handle them with care. When you trim off a damaged frond, cut it off at the base, or the stub will turn brown and look unattractive.

ASPARAGUS FERNS

Botanical name: *Asparagus densiflorus 'Sprengeri'*
Trade name: Sprengeri
Family & Form: Fern (trailing)
Standard size: 10" hanging basket
Other sizes: 6" to 12" basket
Minimum light: 80fc (High artificial)
Minimum ideal light: 150fc+ (North window)
Care: Easiest of all Ferns
Physical durability: Very good
Resistance to pests: Very good
Similar plants: Other Asparagus species

WHEN TO WATER: Plants in minimum light (80fc) in a 10" pot.

>   BY TOUCH: The soil surface gets fairly dry before you water.
>   MOISTURE METER: Insert 3", water when indicator is 1/2 or less.
>   VISUALLY: The needles collapse along the stems.

HOW MUCH TO WATER: Because these Ferns have thick, water-storing roots, you keep them much drier than the Boston Ferns. In light below 150fc, there is always a danger of overwatering, so moisten just the dry soil. In 80fc, a 10" hanging basket requires approximately 2 cups of water every 14 days.

OVERWATERING: The tips of the needles of this family of Ferns turn yellow when the plant is too wet. In extreme cases of overwatering, the needles shed when they are green. The plant will dry itself out rapidly, so you need only skip one watering to cure an Asparagus Fern with wet roots. Thereafter, water it sparingly for the next 4-6 weeks.

UNDERWATERING: When you see masses of bright yellow leaves on this plant, you know the roots are far too dry. When a plant is installed fresh from the greenhouse, the sides of the plant that receive the least light will also turn yellow. To judge underwatering, you must keep your eye on the foliage on the bright-light side and water accordingly.

GROWTH & CARE: There are several common Asparagus Fern varieties. The 'Sprengeri' is the most popular as a hanging basket, while the short and bushy 'Meyeri' makes a good table plant, and the 'Retrofractus' (Ming Fern) sits on the floor and grows nearly upright.

Asparagus Ferns are unique! They grow new foliage by sending out a spear (the stem), that reaches full length before the needles unfold. Trim them back only if they interfere with the décor of the room. Otherwise, thin them out by cutting the runners off at the base. This will encourage new growth to develop on the side that gets the most light. Their only disadvantage is a messy habit of shedding needles. However, they are superior to the Boston Fern family in durability and manageability.

LEATHER FERN

Botanical name: *Rumohra adiantiformis*
Trade name: Rumohra
Family & Form: Fern (bushy)
Standard size: 8" hanging basket
Other sizes: Occasionally to 10" pot
Minimum light: 30fc (Low artificial)
Ideal light: 70fc+ (Avg. artificial)
Care: Extremely easy
Physical durability: Very good
Resistance to pests: Very good
Similar plants: Rabbit's Foot Fern

DESCRIPTION: This is the toughest of all Ferns and it tolerates the least light. The fronds are plastic, like the Wax plant. The only way to damage it is by keeping it too wet. Expect it to growth slowly, even in bright light. It rarely requires repotting. In 30fc it goes 21 days without water. Feel the leaves for moisture before watering this Fern.

BIRD'S NEST

Botanical name: *Asplenium nidus*
Trade name: Asplenium
Family & Form: Fern (rosette)
Standard size: 8" pot
Other sizes: 6" to 12"+ pot
Minimum light: 50fc (Avg. artificial)
Ideal light: 100fc+ (High artificial)
Care: Difficult in low humidity
Physical durability: Poor
Resistance to pests: Fairly good
Similar plants: Some Aroids

DESCRIPTION: The beautiful fronds on this Fern mark-up easily from drafts or even small degrees of miswatering. It's one of the few plants that fares poorly in full sun. Dry it out halfway down between watering (approx. 1½ cups every 14 days in 50fc). It makes a perfect table centerpiece and can live in the same pot for five years.

STAGHORN FERN

Botanical name: *Platycerium species*
Trade name: Platycerium
Family & Form: Fern (bushy)
Standard size: 8" pot
Other sizes: 6" to 12"+ pot
Minimum light: 80fc (High artificial)
Ideal light: 150fc+ (North window)
Care: Easier when small
Physical durability: Fairly good
Resistance to pests: Very good
Similar plants: All Platyceriums

DESCRIPTION: These ferns are prized for their huge antler-shaped leaves. Often they are mounted on a slab of wood and hung on the wall. They like to be dry, so water them when the fronds start to droop. When planted in a pot, they need less water than when they are mounted. A plant in an 8" pot takes approx. 3 cups of water every 14 days in 80fc of light.

TABLE FERN

Botanical name: *Pteris species*
Trade name: Pteris or species name
Family & Form: Fern (bushy)
Standard size: 6" pot
Other sizes: Occasionally 10" pot
Minimum light: 60fc (Avg. artificial)
Ideal light: 100fc+ (High artificial)
Care: Fairly easy
Physical durability: Very good
Resistance to pests: Very good
Similar plants: Other delicate ferns

DESCRIPTION: There is a big variety of foliage colorations in the family of Table Ferns. They grow easily but they crisp-up within a few days when they dry out. They require water 2-3 times a week for best results. Keep them moist except in lower light levels. They are perfect in large planter gardens. Try planting several varieties together for a table arrangement.

PEACE LILIES

Botanical name: *Spathiphyllum species*
Trade name: Spath or species name
Family & Form: Lily (bush)
Standard size: 6" to 8" pot
Other sizes: Occasionally to 14" pot
Minimum light: 40fc (Low artificial)
Minimum ideal light: 60fc+ (Avg. artificial)
Care: Reasonably easy
Physical durability: Very good
Resistance to pests: Fairly good
Similar plants: All Spathiphyllums

WHEN TO WATER: Plants in minimum light (40fc) in an 8" pot.

BY TOUCH: The soil surface gets barely moist before you water.
MOISTURE METER: Insert 3", water when indicator is 1/2 or less.
VISUALLY: The leaves collapse rapidly as their moisture is used.

HOW MUCH TO WATER: In light below 60fc, you must be careful to water only the soil that has dried since the prior time (at least 1/3 down, approx. 2½ cups every 7 days). In medium light, allow to dry at least 1/4 down before watering. In bright window light, keep them fairly moist (approx. 4-5 cups every 7 days in 250fc).

OVERWATERING: When the roots of these plants are too wet for a short time, the tips of the leaves turn yellow. When the roots stay too wet for 6 to 8 weeks, they rot and the plant collapses. With most plants, you withhold water for a week or two when you want to dry them out; however, with the Peace Lily give it half as much water as normal.

UNDERWATERING: This plant is unique because the leaves can completely collapse over the edge of the pot as soon as the soil gets too dry. It can happen in 2 or 3 days. When you water a collapsed plant, the leaves usually stand up within 6-18 hours after you have watered. Brown tips and flowers mean slightly dry roots.

GROWTH & CARE: Spathiphyllums have the ability to grow beautiful white flowers, even in artificial light. The more rootbound they are, the more flowers they produce. They range in size from 1" tall (Wallisii) to 6" tall (Mauna Loa). All of them require the same care, except that the smaller ones are slightly easier. The only danger to Spathiphyllums is hot drafts, which pulls moisture from the leaves faster than the roots can replace it. They are troubled occasionally from spider mites and mealy bugs, but they recover rapidly when sprayed. Because these plants grow wider, rather than taller, they can be divided every 2 or 3 years. This is preferable to putting them in a larger pot, since these plants have a tendency to become extremely bushy.

CAST IRON PLANT

Botanical name: *Aspidistra eliator*
Trade name: Aspidistra
Family & Form: Lily (clump)
Standard size: 10" pot
Other sizes: 6" to 14" pot
Minimum light: 20fc (Low artificial)
Minimum ideal light: 50fc+ (Avg. artificial)
Care: Extremely easy
Physical durability: Excellent
Resistance to pests: Very good
Similar plants: Sansieviera species

WHEN TO WATER: Plants in minimum light (20fc) in a 10" pot.

> BY TOUCH: Water only when there is no moisture at all in the soil.
> MOISTURE METER: The meter will show no moisture in the soil.
> VISUALLY: As the foliage dries, the leaves fold inward.

HOW MUCH TO WATER: When the Aspidistra gets less than 30fc, it can take months to dry out. Therefore, except in a sunny window, you must pour the water sparingly or risk overwatering. The Sword plant is handled similarly, but it requires even less water (approx. 1½ qts. every 3-6 months in 30fc).

OVERWATERING: The Aspidistra and the Sansevieria have thick underground stems that store water during periods of drought. These plants can store almost as much water as a Cactus. When they are overwatered, the tips of the leaves turn yellow and mushy – but the large underground stems also get damaged. Always keep them dry.

UNDERWATERING: Like the Palms and Canes, these two members of the Lily family get brown leaf tips when the roots are overly dry. Because the brownness develops so slowly, you should wait until ⅛" of the tip is brown before you water – especially in low light. The Sansevieria can go six months between waterings without burning noticeably.

GROWTH & CARE: It's unfortunate that the Aspidistra and Sansevieria don't grow larger, because they are certainly the most durable plants you will find. The variegated Aspidistra is quite attractive, and there are numerous colorations in the various varieties of Sansevieria.

Both these plants grow slowly, so don't expect more than 20% new growth every year. They both grow straight up, out of the pot, so they are perfect for a low light, confined corner. When they are fresh from the nursery, they may need water every 2 or 3 weeks for the first 3 months, and even more often in a sunny window. The Aspidistra is the more delicate of the two, and burns faster – but the Sansevieria can die rapidly from a single overwatering. Propagate by dividing them into clumps.

JADE PLANT

Botanical name: *Crassula argentea*
Trade name: Crassula or species name
Family & Form: Succulent (tree-like)
Standard size: 2'-3' tall
Other sizes: Occasionally 6'-8'
Minimum light: 80fc (High artificial)
Minimum ideal light: 250fc+ (East window)
Care: Easy when kept dry
Physical durability: Very good
Resistance to pests: Very good
Similar plants: Most succulents

WHEN TO WATER: Plants in minimum light (80fc) in a 10" pot.

BY TOUCH: Feel the leaves, water when they are flabby.
MOISTURE METER: Insert 7", water when needle says 1/4 or less.
VISUALLY: The leaves get noticeably wrinkled.

HOW MUCH TO WATER: All Succulents are watered the same way. Wait until the foliage has changed from the plumpness it assumes after being watered – to the softness that comes when the internal stores of moisture have been used. A 2½' tall Jade requires water 3-4 weeks in 100fc.

OVERWATERING: Branching Succulents like the Jade react similar to Trees when the roots are too wet – the leaves fall off when they are still green. In extreme cases of overwatering, whole branches can fall off. When you have overwatered a Succulent, all you can do is hope it dries out before it sheds all its foliage. It may take 2-3 months to dry.

UNDERWATERING: Even after a Succulent has been bone dry for a month or more, you will notice that only a few of the leaves have shriveled up beyond recovery. A good watering recovers the plant within 24 hours. The moral with Succulents is not to water them until they are obviously dried out. Then, like Cacti, they are nearly impossible to kill.

GROWTH & CARE: No doubt you have seen ancient Jade plants, 10 years old or more. When they grow in light that is less than full sun, the branches get leggy, and most of the leaves are congregated at the ends of the branches. Pruning them back every two years helps make them more compact and bushier. A Jade grown in a sunny window should be pinched back – to shape the plant and to prevent leggy growth. The Jade is typical of Succulents that have leaves, branches, and a main trunk or stem. Other kinds grow like the Aloe Vera (leaves growing around a short central stalk, usually forming a rosette). The later types are more durable since the individual leaves are larger, and the central stalk gives them more stability.

CACTI

Cacti are similar to succulents. They prefer to have lots of light and relatively little water. Because they store so much water in their bodies, the soil can be bone dry for a year or more before they start to shrivel. This means they can stay in a dark spot for an extended period.

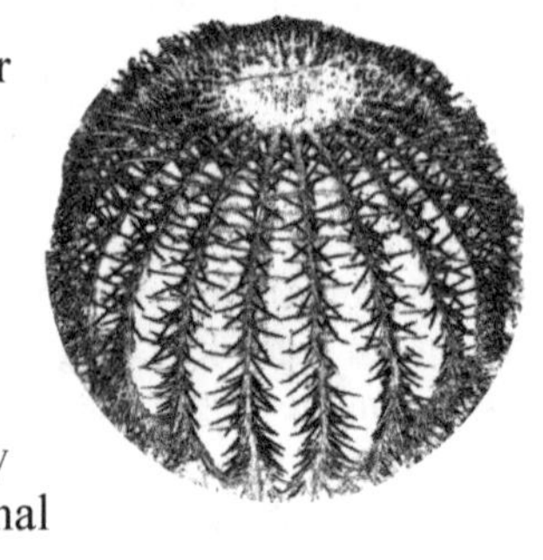

Since Cacti grow slowly, they are accustomed to manufacturing food slowly. In a dark spot they grow imperceptibly, and thus retain their original appearance.

The large, barrel-types can easily last two years or more in the dark – providing they are not watered. Even in bright window light, a 2' tall Barrel Cactus will need no more than two waterings a year. Cacti are like the rest of the tropicals in showing overwatering – their surface color gets a paler mottled green.

Tall Peruvian Cacti are not quite as durable – since they grow relatively fast for a cactus. They still last 6-12 months in low light (40fc or more). They prefer to have at least 150fc, but 300fc is better. A 6' Peruvian cactus, growing in 300fc, needs water every 3-4 months.

One of the most beautiful family members is the Golden Barrel. It is large and fat and the bright yellow color is exciting in any indoor location. It has a level of durability that rivals the other barrel cacti. Opuntias (Rabbit Ear Cacti) are another exciting group which exhibits a wide range of coloration and shape. They are equivalent in care to the sturdiest of the succulents (agaves and aloes). The 'ears' are easy to propagate by callousing them and sticking them in a pot of dry soil. Wait 3-6 months before watering. When an Opuntia is growing in 300fc of light, it needs water every one or two months during the summer.

Most cacti need plenty of light to flower (over 1000fc), but the Crown of Thorns (*Eurphorbia spendens*), will produce a few flowers in light as low as 200fc. The tiny, blood-red blooms contrast refreshingly with the tiny, oval leaves. A Crown of Thorns needs more water than the more traditionally-shaped cacti. In bright window light (500fc+), you may have to water every two or three weeks.

Like succulents, you water cacti by feeling the flesh for moisture content. Wait until the skin is wrinkled and soft before pouring the water. A single dose of water, before the plant is ready to accept it, often spells doom indoors.

Some of the smaller cacti look good when planted in a dish garden with miniatre succulents. You will have to water the plants separately, since the succulents need water more often.

Cacti are not used that often indoors. They are dangerous because of sharp spines and large cacti are more expensive than other tropicals of equivalent height. Cacti are effective when planted in a bed with succulents which often produce flowers in 300fc of light. The succulents provide a fullness that takes away from the starkness of mature cacti.